Adil BARRA

Modeling, identification and control of physical and logistical systems

Adil BARRA

Modeling, identification and control of physical and logistical systems

ScienciaScripts

Imprint

Cover image: www.ingimage.com

This book is a translation from the original published under ISBN 978-620-6-72793-4.

Publisher:
Sciencia Scripts
is a trademark of
Dodo Books Indian Ocean Ltd. and OmniScriptum S.R.L publishing group

120 High Road, East Finchley, London, N2 9ED, United Kingdom
Str. Armeneasca 28/1, office 1, Chisinau MD-2012, Republic of Moldova, Europe
Printed at: see last page
ISBN: 978-620-8-33113-9

Table of contents

Foreword

This book presents a summary of the author's research activities, focusing on the modelling, identification and control of physical and logistical systems. These activities are divided into three main areas:

The first part focuses on the modelling and control of a wind turbine based on a MADA dual-feed asynchronous machine, while considering the hysteresis behaviour of its magnetic circuit. The second part targets the identification of the internal parameters of asynchronous machines using Kalman filters. The aim of this study is to locate the variation in parameters in order to diagnose the probability of breakdowns.

The second axis is essentially concerned with the feasibility study of proposals for matrices constituting the singular representation of physical systems with delay. The aim of this study is to propose models that are as close as possible to real systems.

The third axis focuses on the regulation of the stock level of a warehouse, the essential objective of this study being to determine the instant of replenishment in order to optimise the preparation and supply times of the storage spaces, the number of trips and the storage space.

Preface

The present document, presents a synthesis of my research activities accentuated around the modeling, the identification and the control of the physical and logistic systems, and which one can classify in three main axes:

The first axis is divided into two parts, where the first focuses on the modeling and control of a wind turbine based on a double-feed induction generator DFIG, while considering the hysteresis behavior of its magnetic circuit. The second part is dedicated to the identification of the internal parameters of the asynchronous machines by the Kalman filters theory, this study aims to locate the variation of the parameters in order to diagnose the probability of breakdowns.

The second axis is mainly interested in the feasibility study of the proposals of the matrices constituting the singular representation of physical systems with delays, this study aims to propose models which approach the most to the real systems behavior.

The third axis focuses on the regulation of a warehouse stock level, the essential objective of this study is to determine the moment of lunching the replenishments order to optimise the times of preparation and the supply spaces of storage, number of trips.

General introduction

During this period, I became interested in Smart Grids, where the main idea was to set up a system for managing a network's electricity supply sources, while favouring the cheapest source. In addition to the electricity grid, this system will combine renewable energies, which are generally wind and photovoltaic, as well as fuel cell cells as a means of producing and storing hydrogen, a generator and batteries. The system will be managed by an objective cost function, where several scenarios will be implemented depending on a number of meteorological and financial parameters.

This project began with wind power systems, where I proposed a new modelling and control system that takes account of magnetic saturation. This brings the proposed model even closer to the real system and requires less switching on the converter triggers. To validate this proposal, a comparison is carried out to prove the superiority of the proposed control law over conventional controls. For better management of the wind power system within a smart grid, we need to diagnose the state of health of the machine in real time. This is the context of our second work, which focuses on identifying the internal parameters of the asynchronous machine, The variation of these parameters can provide us with information in real time on the state of the machine so that we can generate alarms or preventive maintenance actions, all this without integrating physical sensors which reduce the reliability of the machine, while requiring periodic control and calibration, and even replacement in the event of wear.

Mathematical models of physical systems are generally represented by cancelling the singular matrix in the state representation. This simplification generates a mismatch between the physical system and its mathematical representation. This is the subject of our third project, which focuses on the singular representation of physical systems, while adding delays that may be caused by the sensor, data transmission time, processing time, etc.

Joining the GEITIIL laboratory enabled me to get involved in subjects linked to the logistics field, where I brought my touch as an automatician to the management of logistics systems. This gave rise to work on modelling and controlling the stock level in a warehouse, where the aim is to automatically generate the times at which replenishment is launched, which are the most optimal in terms of cost and time.

At the same time, four other thesis topics were launched last year to complete the team, which will focus primarily on carrying out research work on the smart grid project. In what follows, this document will be divided into three sections:

The first axis is divided into two parts, the first of which concerns the modelling and control of a wind turbine based on a MADA dual-feed asynchronous machine, taking into account the hysteresis behaviour of its magnetic circuit. The second part is devoted to the identification of the internal parameters of the asynchronous motor using the Kalman filter. The aim of this study is to locate the variation in the parameters in order to diagnose the probability of breakdowns.

The second axis is essentially concerned with the feasibility study of matrices constituting the singular representation of physical systems with delay. The aim of this study is to propose models that are as close as possible to real systems.

The third axis focuses on the regulation of stock levels in a warehouse. The main objective of this study is to determine the times at which replenishments should be launched in order to optimise preparation times, the supply of storage shelves, the number of movements and the main storage space.

The work carried out by our team has been published in indexed international journals and presented at international conferences.

Axe 1. Identification, modelling and control of electrical machines

Chapter I: Comparative study between optimal control of DFIG flux and velocity in the presence of magnetic hysteresis and conventional control strategies for wind energy systems

Summary of Chapter I:

In the field of wind energy systems, the control objective most commonly addressed by authors is the control of the speed and flux of the double-fed induction generator (DFIG). For reasons of computational complexity, the mathematical representation of the generator is generally modelled assuming a linear magnetic characteristic. Based on this assumption, the rotor flux control will aim to follow a fixed flux reference, mainly its nominal value. The disadvantage of this strategy is that, in practice, by neglecting the non-linearity of the generator's magnetic characteristic, optimum performance control cannot be achieved.

In this study, by considering the hysteresis of the magnetic flux and the saturation characteristic, a new modelling of the DFIG is presented. In addition, a new Backstepping controller with optimal speed and flux is designed based on Lyapunov theory. To prove its performance, a comparative study with different control strategies will be carried out. Simulations taking into account a wide range of wind speed variations are carried out in the Matlab/Simulink environment.

I.1.1 Introduction to the chapter

In recent decades, renewable energies, including wind turbines, have attracted major attention worldwide. Regardless of their epic performance, DFIG-based wind turbines are complex and their control law design is challenging [7], as the model considered is multi-variable and highly non-linear.

DFIG control using a linear or non-linear controller has been the subject of several previous studies [6-10]. The objective of speed tracking and flux regulation for MPPT is the most discussed topic present in the literature. However, the proposed controllers work as expected when a linear magnetic characteristic is assumed. In practice, this assumption is not realistic [3, 4, 5], as the DFIG magnetic characteristic is non-linear and exhibits errors and adverse oscillations [15] due to hysteresis and saturation produced by the magnetic material. However, if rotor flux regulation is performed close to a constant nominal value, standard models (ignoring magnetic non-linearity) can still be used in the speed control design. It is only when the aerodynamic torque is close to the nominal value of the DFIG electromagnetic torque that the DFIG efficiency reaches its maximum. However, the aerodynamic torque is generally not fixed a priori and can be subject to large variations in real applications, due to the considerable variation in wind speed.

To solve the above problems, a new speed control system has been developed, using on-line adjustment of the rotor flux reference to track the optimum speed reference in the presence of significant variation in wind speed. The optimum flux reference will also experience considerable range variations in these circumstances, meaning considerable deviations from the operating point on the magnetic characteristic. Therefore, the controller design must be based on a model that takes into account the non-linearity of the machine's magnetic circuit in order to provide high control performance whatever the DFIG's operating mode. Both hysteresis and saturation characterise this non-linearity.

To our knowledge, no scientific work has yet dealt with MPPT control of the DFIG under saturation/hysteresis phenomena. In [12]-[13], non-linear controllers were developed for a wind power system taking into account only magnetic saturation (neglecting the hysteresis effect). This assumption makes it relatively easy to develop the DFIG model and to synthesise the MPPT controller. However, in practice, hysteresis degrades the expected performance of the wind system (accuracy and oscillation problems).

Numerous mathematical hysteresis models have been developed to describe hysteresis phenomena. They can be classified as follows:

a) Operator-based hysteresis models such as the Preisach model and the Prandtl-Ishlinskii model, where an infinite number of hysteretic operators are involved in their integral model, implying high computational complexities [3].

b) Hysteresis models based on differential equations such as the Backlash model, the Bouc-Wen model or the Duhem model, where the finite number of hysteretic operators can be easily extended to continuous inputs using approximation and a limiting process, avoiding computational complexity [3].

In the present work, a special case of Duhem's model, the Coleman-hudgdon model [2], is considered to describe the nonlinear magnetic characteristic. This model has proved very practical for modelling hysteresis in ferromagnetic materials [2, 14].

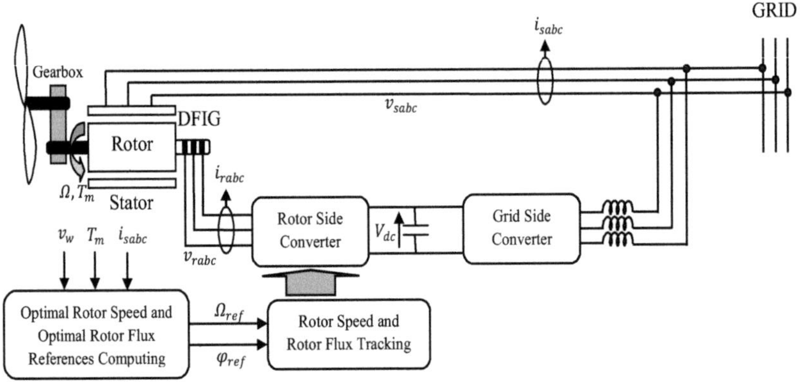

Fig.1. General diagram of the wind energy system under consideration

The non-linearity of the magnetic characteristic will be taken into account when a revised model for the DFIG is constructed in this work. In addition, a new controller for the considered wind system is developed based on this new model. The two control objectives are: i) tracking the optimal rotor speed reference of the DFIG (for MPPT purposes); ii) tracking the optimal rotor flux reference of the DFIG.

Optimising rotor flux involves minimising stator current, which will reduce stator joule losses. The DFIG speed/flux controller created using the backstepping technique proves to be very different from conventional DFIG control strategies that consider constant flux references and assume a linear magnetic characteristic. Despite variations in wind speed, it is formally demonstrated that the newly developed controller is globally asymptotically stable and applies speed and flux to exactly trace their fluctuating reference trajectory.

To demonstrate the value of taking saturation/hysteresis phenomena into account in the design of the controller, the performance of a conventional controller (neglecting saturation/hysteresis) will be evaluated by simulation, compared with that of the proposed new controller. In addition, to verify the robustness of the proposed wind power system, the performance of the controller will be inspected in the presence of various faults in the electrical network (voltage dips, frequency deviations and load variations).

This chapter is organised as follows: Section 2 describes the characterisation of magnetic hysteresis in induction machines. Section 3 describes the model of the induction machine; the speed/flow controller of the machine is designed and analysed in Section 4; the performance of the controller is illustrated by simulations in Section 5.

I.1.2 Characterisation of magnetic saturation hysteresis in asynchronous machines

The Coleman-Hodgdon model is derived from the Duhem model, which was proposed for active hysteresis from 1897 [2]. Using a phenomenological approach, this differential equation hysteresis model focuses on the fact that the output can only change character when the input changes direction. In this section, the Coleman-Hodgdon model and its properties are briefly introduced [1].

A differential hysteresis model can be represented by the Duhem hysteresis model. The Duhem magnetic hysteresis model has been studied extensively by Coleman and Hodgdon. For ease of use, this type of Duhem model will be referred to as the CH model throughout this study [1]. It can be represented in terms of the CH model as :

$$\frac{d\varphi_\mu}{dt} = \frac{1}{2}\left(g_1(i_\mu,\varphi_\mu) - g_2(i_\mu,\varphi_\mu)\right)\left|\frac{di_\mu}{dt}\right| + \frac{1}{2}\left(g_1(i_\mu,\varphi_\mu) - g_2(i_\mu,\varphi_\mu)\right)\frac{di_\mu}{dt}, \forall t \in (0,T) \quad (I.1)$$

$$\varphi_\mu(0) = \varphi_{\mu 0} \quad (I.2)$$

$$g_1 = g(i_\mu) + \alpha(f(i_\mu) - \varphi_\mu) \quad (I.3)$$

$$g_2 = g(i_\mu) - \alpha(f(i_\mu) - \varphi_\mu) \quad (I.4)$$

By introducing (3), (4) into (1), we find:

$$\frac{d\varphi_\mu}{dt} = \alpha\left|\frac{di_\mu}{dt}\right|[f(i_\mu) - \varphi_\mu] + \frac{di_\mu}{dt}g(i_\mu) \quad (I.5)$$

where$\alpha$is a positive constant, and all three conditions must be satisfied.

- ***Condition 1:*** *f*(.) is piecewise continuous, monotonically increasing, with $\lim_{i_\mu \to \infty} f'(i_\mu)$ finite;
- ***Condition 2:*** g(.) is piecewise continuous, with $\lim_{i_\mu \to \infty} g(i_\mu) = \lim_{i_\mu \to \infty} f'(i_\mu)$
- ***Condition 3***: For all $i_\mu > 0,\ f'(i_\mu) > g(i_\mu) > \alpha e^{\alpha i_\mu} \int_{i_\mu}^{\infty} |f'(\zeta) - g(\zeta)| e^{-\alpha\zeta}\, d\zeta$

To illustrate the case studied previously for the generator considered in this work, the corresponding functions f(.),g(.) of the C-H model are chosen as follows:

$$f(i_\mu) = c \tanh(i_\mu) + a i_\mu \qquad \text{(I.6)}$$

$$g(i_\mu) = f'(i_\mu)(1 - b e^{-|i_\mu|}) \qquad \text{(I.7)}$$

Note that the functions f(.) and g(.) satisfy the three conditions presented above. Experimental measurements of the magnetic characteristic can be used to calculate the parameters c and α. The shape of the hysteresis cycle obtained is shown in Fig.2.

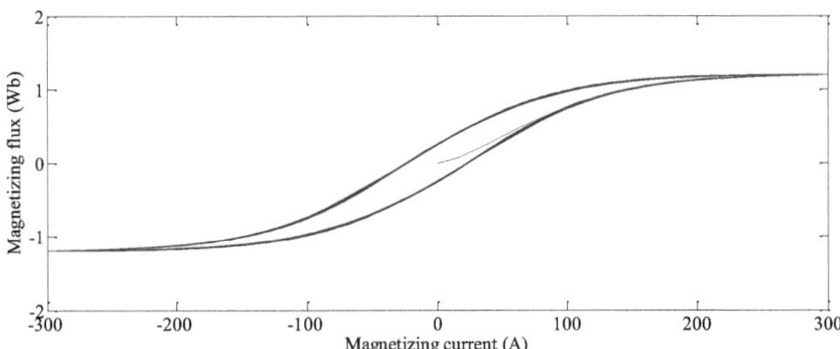

Fig.2. Hysteresis curve given by the C-H model for the 1st phase.

I.1.3 Double-fed asynchronous generator model

A. Electrical and flux equations for alternating current machines

The dual-fed induction generator can be modelled by its electrical and mechanical equations. A stationary reference frame (α, β) will be used to develop the control model given in this study. The equations of the AC electric machine are given by :

- Electrical equations of the stator [9]

$$V_{s\alpha} = R_s i_{s\alpha} + \frac{d\varphi_{s\alpha}}{dt} \quad \text{(I.8)}$$

$$V_{s\beta} = R_s i_{s\beta} + \frac{d\varphi_{s\beta}}{dt} \quad \text{(I.9)}$$

- Electrical equations of the rotor [9]

$$V_{r\alpha} = R_r i_{r\alpha} + \frac{d\varphi_{r\alpha}}{dt} + \omega\varphi_{r\beta} \quad \text{(I.10)}$$

$$V_{r\beta} = R_r i_{r\beta} + \frac{d\varphi_{r\beta}}{dt} - \omega\varphi_{r\alpha} \quad \text{(I.11)}$$

Similarly, the components (α,β) of the stator and rotor fluxes verify [9] :

$$\varphi_{s\alpha} = L_s i_{s\alpha} + \varphi_{\mu\alpha} \quad \text{(I.12)}$$

$$\varphi_{s\beta} = L_s i_{s\beta} + \varphi_{\mu\beta} \quad \text{(I.13)}$$

$$\varphi_{r\alpha} = \varphi_{\mu\alpha} \quad \text{(I.14)}$$

$$\varphi_{r\beta} = \varphi_{\mu\beta} \quad \text{(I.15)}$$

the components (α, β) of the magnetising currents satisfy :$i_{r\alpha} = i_{\mu\alpha} - i_{s\alpha}$ (I.16)

$$i_{r\beta} = i_{\mu\beta} - i_{s\beta} \quad \text{(I.17)}$$

The classic DFIG state space model is generally presented with this vector

$X = \left[i_{s\alpha}, i_{s\beta}, \varphi_{r\alpha}, \varphi_{r\beta}, \Omega\right]^T$ [11], but to introduce the hysteresis character into the state-space model, the vector X will be augmented by the components (α, β) of the magnetising current i_μ. The new state vector considered is given by :

$$X = \left[i_{s\alpha}, i_{s\beta}, \varphi_{r\alpha}, \varphi_{r\beta}, \Omega, i_{\mu\alpha}, i_{\mu\beta}\right]^T \quad \text{(I.18)}$$

B. Flux state equations of the rotor

The flux dynamics in the rotor are given by equations (10) - (11). However, they involve the components (α, β) of the rotor current. These are not considered as state variables (see 18). We will try to express them in terms of the components of the vector X.

Replacing (16)-(17) in (10)-(11) gives the equations of state for the rotor fluxes.

$$\frac{d\varphi_{r\alpha}}{dt} = V_{r\alpha} - R_r i_{\mu\alpha} + R_r i_{s\alpha} - \omega\varphi_{r\beta} \quad (I.19)$$

$$\frac{d\varphi_{r\beta}}{dt} = V_{r\beta} - R_r i_{\mu\beta} + R_r i_{s\beta} + \omega\varphi_{r\alpha} \quad (I.20)$$

C. Current equations for the stator :

By introducing (12) - (13) into (8) - (9), we deduce the dynamics of the currents in the stator:

$$\frac{di_{s\alpha}}{dt} = -\frac{R_s}{L_s} i_{s\alpha} + \frac{1}{L_s} V_{s\alpha} - \frac{1}{L_s}\frac{d\varphi_{\mu\alpha}}{dt} \quad (I.21)$$

$$\frac{di_{s\beta}}{dt} = -\frac{R_s}{L_s} i_{s\beta} + \frac{1}{L_s} V_{s\beta} - \frac{1}{L_s}\frac{d\varphi_{\mu\beta}}{dt} \quad (I.22)$$

As the components (α, β) of the magnetic flux are not considered as state variables, then, introducing (14) - (15), (16) - (17) into (19) - (20), we obtain:

$$\frac{di_{s\alpha}}{dt} = -\frac{R_s}{L_s} i_{s\alpha} + \frac{1}{L_s} V_{s\alpha} - \frac{1}{L_s}\left[V_{r\alpha} - R_r i_{\mu\alpha} + R_r i_{s\alpha} - \omega\varphi_{r\beta}\right] \quad (I.23)$$

$$\frac{di_{s\beta}}{dt} = -\frac{R_s}{L_s} i_{s\beta} + \frac{1}{L_s} V_{s\beta} - \frac{1}{L_s}\left[V_{r\beta} - R_r i_{\mu\beta} + R_r i_{s\beta} + \omega\varphi_{r\alpha}\right] \quad (I.24)$$

C. Rotor speed equations of state :

The electromagnetic torque T_m delivered by the generator is given by [9] :

$$T_m = p\left(\varphi_{s\alpha} i_{s\beta} - \varphi_{s\beta} i_{s\alpha}\right) \quad (I.25)$$

Using the flux equations (12) - (13), the electromagnetic torque can be expressed as a function of the components (α, β) of the rotor flux (considered as state variables), namely:

$$T_m = p\left(\varphi_{r\alpha} i_{s\beta} - \varphi_{r\beta} i_{s\alpha}\right) \quad (I.26)$$

By applying the principle of rotational dynamics, we can derive the equation of state for the rotor speed

$$\frac{d\Omega}{dt} = \frac{p}{J}\left(\varphi_{r\alpha} i_{s\beta} - \varphi_{r\beta} i_{s\alpha}\right) - \frac{T_L}{J} - \frac{f}{J}\Omega \tag{I.27}$$

D. Equations of state of the magnetising current

Hysteresis is taken into account by the C-H model defined in (5). This expressed the dynamics of the magnetising flux. For convenience, this model is used again in this paragraph.

$$\frac{d\varphi_{\mu k}}{dt} = \frac{di_{\mu k}}{dt}\left(\alpha sign(\dot{i}_{\mu k})\left[f(i_{\mu k}) - \varphi_{\mu k}\right] + g(i_{\mu k})\right) \tag{I.28}$$

where k=1,2,3 represents the different phases. Otherwise, it is easy to check (by referring to the hysteresis cycle Fig.2) that the direction of variation of the current $i_{\mu k}$ is identical to the direction of flux variation $\varphi_{\mu k}$. Consequently, we have

$$sign(\dot{i}_{\mu k}) = sign(\dot{\varphi}_{\mu k}) \tag{I.29}$$

Let's define h as the quantity

$$h(i_{\mu k}, \varphi_{\mu k}, \dot{i}_{\mu k}) = \frac{1}{\left(\alpha sign(\dot{i}_{\mu k})[f(i_{\mu k}) - \varphi_{\mu k}] + g(i_{\mu k})\right)} \tag{I.30}$$

where

$$\frac{di_{\mu k}}{dt} = h(i_{\mu k}, \varphi_{\mu k}, \dot{i}_{\mu k})\frac{d\varphi_{\mu k}}{dt} \tag{I.31}$$

The application of the direct and inverse Concordia transformation to the three-phase system defined by (31), implies:

$$\frac{d}{dt}\begin{bmatrix} i_{\mu\alpha} \\ i_{\mu\beta} \end{bmatrix} = C_{23}\begin{bmatrix} h_1 & 0 & 0 \\ 0 & h_2 & 0 \\ 0 & 0 & h_3 \end{bmatrix} * C_{32}\frac{d}{dt}\begin{bmatrix} \varphi_{\mu\alpha} \\ \varphi_{\mu\beta} \end{bmatrix} \tag{I.32}$$

Where $h_k$can be defined as

$$h_k = h(i_{\mu k}, \varphi_{\mu k}, \dot{i}_{\mu k}) \tag{I.33}$$

By substituting (14), (15), (19) and (20) into the vector equation (32), the following two sixth and seventh state space equations of the AC machine are constructed:

$$\frac{di_{\mu\alpha}}{dt} = \frac{1}{6}(4h_1 + h_2 + h_3)(V_{r\alpha} - R_r i_{\mu\alpha} + R_r i_{s\alpha} - \omega\varphi_{r\beta}) - \frac{\sqrt{3}}{6}(h_2 - h_3)(V_{r\beta} - R_r i_{\mu\beta} + R_r i_{s\beta} - \omega\varphi_{r\alpha}) \quad (I.34)$$

$$\frac{di_{\mu\beta}}{dt} = -\frac{\sqrt{3}}{6}(h_2 - h_3)(V_{r\alpha} - R_r i_{\mu\alpha} + R_r i_{s\alpha} - \omega\varphi_{r\beta}) + \frac{1}{2}(h_2 + h_3)(V_{r\beta} - R_r i_{\mu\beta} + R_r i_{s\beta} - \omega\varphi_{r\alpha}) \quad (I.35)$$

Equations (19), (20), (23), (24), (27), (34) and (35) represent the established DFIG state space model. For a more compact form, consider the following state space model:

$$\dot{X} = F(X) + Gu \quad (I.36)$$

$$y = \lambda(X) = \begin{bmatrix} \Omega \\ \varphi_{r\alpha}^2 + \varphi_{r\beta}^2 \end{bmatrix} \quad (I.37)$$

With :

$$X = [i_{s\alpha}, i_{s\beta}, \varphi_{r\alpha}, \varphi_{r\beta}, \Omega, i_{\mu\alpha}, i_{\mu\beta}]^T \quad (I.38)$$

$$u = [V_{r\alpha}, V_{r\beta}]^T \quad (I.39)$$

$$G = \begin{bmatrix} -\frac{1}{L_s} & 0 & 1 & 0 & 0 & 0 & 0 \\ 0 & -\frac{1}{L_s} & 0 & 1 & 0 & 0 & 0 \end{bmatrix}^T \quad (I.40)$$

$$F(X) = \begin{bmatrix} -\frac{R_s}{L_s} i_{s\alpha} + \frac{1}{L_s} V_{s\alpha} - \frac{1}{L_s}[-R_r i_{\mu\alpha} + R_r i_{s\alpha} - \omega\varphi_{r\beta}] \\ -\frac{R_s}{L_s} i_{s\beta} + \frac{1}{L_s} V_{s\beta} - \frac{1}{L_s}[-R_r i_{\mu\beta} + R_r i_{s\beta} + \omega\varphi_{r\alpha}] \\ -R_r i_{\mu\alpha} + R_r i_{s\alpha} - \omega\varphi_{r\beta} \\ -R_r i_{\mu\beta} + R_r i_{s\beta} + \omega\varphi_{r\alpha} \\ \frac{P}{J}(\varphi_{r\alpha} i_{s\beta} - \varphi_{r\beta} i_{s\alpha}) - \frac{T_L}{J} \\ f_6(X) \\ f_7(X) \end{bmatrix} \quad (I.41)$$

With :

$$f_6(X) = \frac{1}{6}(4h_1 + h_2 + h_3)(V_{r\alpha} - R_r i_{\mu\alpha} + R_r i_{s\alpha} - \omega\varphi_{r\beta}) - \frac{\sqrt{3}}{6}(h_2 - h_3)(V_{r\beta} - R_r i_{\mu\beta} + R_r i_{s\beta} - \omega\varphi_{r\alpha}) \quad (I.42)$$

$$f_7(X) = -\frac{\sqrt{3}}{6}(h_2 - h_3)(V_{r\alpha} - R_r i_{\mu\alpha} + R_r i_{s\alpha} - \omega\varphi_{r\beta}) + \frac{1}{2}(h_2 + h_3)(V_{r\beta} - R_r i_{\mu\beta} + R_r i_{s\beta} - \omega\varphi_{r\alpha}) \quad (I.43)$$

I.1.4 Design and analysis of the speed and flow controller

There are two main elements to any DFIG control strategy:

i) The rotor flux reference generator used

ii) The DFIG model considered when designing the controller.

In this context, depending on how these components are designed, four control strategies are envisaged (see Table 2).

Table 4. Different speed control strategies.

		Reference Flux generator	
		Nominal Flow	Optimal Flux
Magnetic Characteristic model	Linear	*LMC-NF*	*LMC-OF*
	Hysteretic	*HMC-NF*	*HMC-OF*

In the existing literature, the most popular DFIG control strategy is the LMC-NF, which is characterised by a non-optimal constant flux reference and an MPPT controller based on the linear magnetic model. Here, the constant flux reference must normally receive the nominal value of the machine flux. The resulting control performance is not satisfactory from an energy point of view, particularly at low wind speeds.

The LMC-OF control strategy is characterised by a variable flux reference, which is calculated by considering the minimum stator current for each torque value. The main advantage of this configuration is to minimise stator joule losses, which will enable more power to be extracted from the generator. The speed regulator is designed taking into account a linear hysteretic magnetic characteristic.

The HMC-NF control strategy (see Table 2) is characterised by a constant flux setpoint. The speed controller is designed taking into account the hysteretic magnetic characteristic. Such a strategy is practically useless as no advantage is gained from the complexity of the model if the flux reference is kept constant.

This chapter focuses on the new control strategy (HMC-OF) which consists in designing an optimal flux reference generator and an MPPT controller (Fig.5). The latter is designed on the basis of the non-linear model (36)-(43) which takes into account the hysteretic characteristic of the DFIG magnetic characteristic. The optimality of the flux reference is supposed to guarantee the minimisation of the stator current required to produce the maximum electromagnetic torque corresponding to a given wind speed. The proposed control strategy is presented in detail in subsections 4.1 and 4.2.

A . Optimum speed reference and flow reference generator:

a) Calculation of optimal flow reference

The design of the optimal flux reference algorithm consists of presenting the optimal stator current as a function of the rotor flux for each value of mechanical torque. For this purpose, representing the mechanical torque in a d-q reference frame seems more interesting from a computing point of view.

If the d component of the reference frame is aligned with the rotor flux, the q component of the flux is zero and all the state variables are constant in steady state.
According to $\varphi_{rq} = 0$ the torque equation becomes

$$T_m = p\varphi_r i_{sq} \tag{I.44}$$

from (23)-(24), the d-q components of the steady-state stator voltages are given by

$$V_{sd} = R_s i_{sd} - \omega_s \varphi_{sq} \quad \text{(I.45}$$

$$V_{sq} = R_s i_{sq} + \omega_s \varphi_{sd} \tag{I.46}$$

Replacing (12)-(13) in (45)-(46), we have:

$$V_{sd} = R_s i_{sd} - l_s \omega_s i_{sq} \quad (I.47)$$

$$V_{sq} = R_s i_{sq} + \omega_s (l_s i_{sd} + \varphi_r) \quad (I.48)$$

Direct connection of the stator to the mains leads to the following equality :

$$V_{smax}^2 = V_{sd}^2 + V_{sq}^2 \quad (I.49)$$

Replacing (47) and (48) in (49) gives :

$$V_{smax}^2 = [R_s^2 + l_s^2 \omega_s^2] i_{sd}^2 + 2\omega_s l_s \varphi_r i_{sd} + [l_s^2 \omega_s^2 + R_s^2] i_{sq}^2 + 2R_s \omega_s \varphi_r i_{sq} + \omega_s^2 \varphi_r^2 \quad (I.50)$$

By introducing (44) into (50), the d component of the stator current verifies

$$i_{sd} = -\frac{\omega_s l_s \varphi_r}{[R_s^2 + l_s^2 \omega_s^2]} + \sqrt{\left[\frac{\omega_s l_s \varphi_r}{[R_s^2 + l_s^2 \omega_s^2]}\right]^2 + \frac{V_{smax}^2}{[R_s^2 + l_s^2 \omega_s^2]} - \left(\frac{T_m}{p\varphi_r}\right)^2 - \frac{2R_s \omega_s \varphi_r}{[R_s^2 + l_s^2 \omega_s^2]} \frac{T_m}{p\varphi_r} - \frac{\omega_s^2}{[R_s^2 + l_s^2 \omega_s^2]} \varphi_r^2} \quad (I.51)$$

Let's consider I_s the stator current standard, then$I_s = \sqrt{{i_{sd}}^2 + {i_{sq}}^2}$ (I.52)

Using (50) and (58), the resulting stator current norm could be given by

$$I_s = \sqrt{\left[-\frac{\omega_s l_s \varphi_r}{[R_s^2 + l_s^2 \omega_s^2]} + \sqrt{\left[\frac{\omega_s l_s \varphi_r}{[R_s^2 + l_s^2 \omega_s^2]}\right]^2 + \frac{V_{smax}^2}{[R_s^2 + l_s^2 \omega_s^2]} - \left(\frac{T_m}{p\varphi_r}\right)^2 - \frac{2R_s \omega_s \varphi_r}{[R_s^2 + l_s^2 \omega_s^2]} \frac{T_m}{p\varphi_r} - \frac{\omega_s^2}{[R_s^2 + l_s^2 \omega_s^2]} \varphi_r^2}\right]^2 +} \sqrt{\left[\frac{T_m}{p\varphi_r}\right]^2} \quad (I.53)$$

Equation (53) expresses I_s as a function of $T_m$and φ_r.

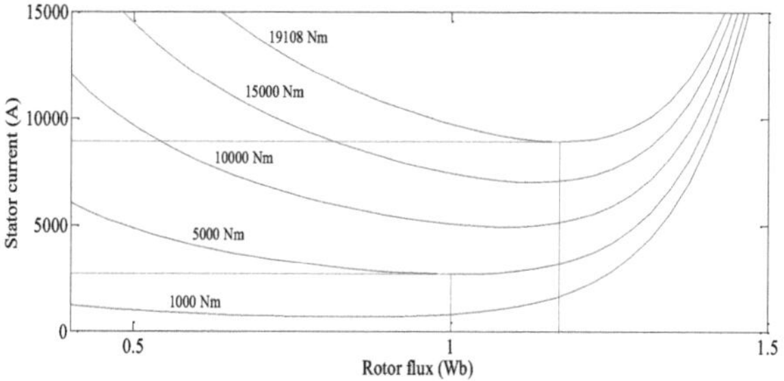

Fig.3. Torque curves (19108, 15000, 10000, 5000 and 1000Nm) each curve designates a minimum current value corresponding to the optimum flux.

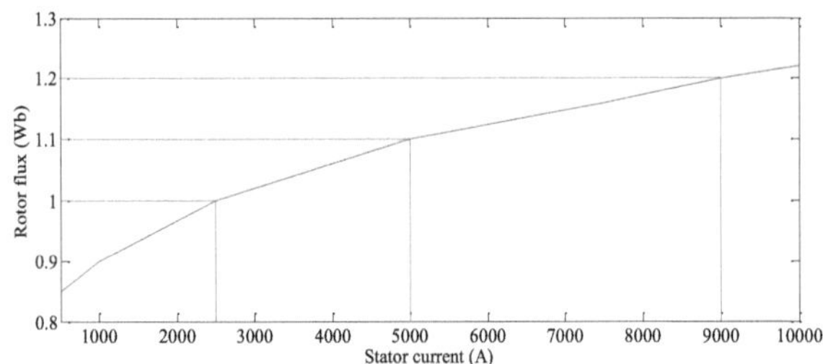

Fig.4.Optimum current-flow characteristic (OCF)

A sample of 9 relevant torque values T_{m_j} (j=1,...,9) has been selected a priori and the corresponding global minima$(\varphi^*_{T_{m_j}}, I^*_{T_{m_j}})$ can be determined graphically as shown in Fig.4 where all curves correspond to the induction machine characterised by the numerical parameters in Table 3. In doing so, a set of 9 global minima$(\varphi^*_{T_{m_j}}, I^*_{T_{m_j}})$ (j=1,...,9) was obtained. The set of minima points was fitted in the least squares sense by a polynomial function of order n, denoted F(.). The degree n=5 proved appropriate for the data under consideration. The resulting polynomial is denoted :

$$F(I_s) = \theta_n I_s^n + \theta_{n-1} I_s^{n-1} + \cdots + \theta_1 I_s + \theta_0 \quad (I.54)$$

Table.5. Coefficientsof the polynomial F(.)

Index	Value	Index	Value	Index	Value
θ_0	0.85	θ_1	−0.0612	θ_2	0.4986
θ_3	−1.719	θ_4	3.1370	θ_5	−2.8389

b) Optimum speed reference :

The main advantage of working with a variable speed reference is that the choice of reference can be optimised to extract the maximum available power. In addition, it guarantees optimum aerodynamic efficiency.

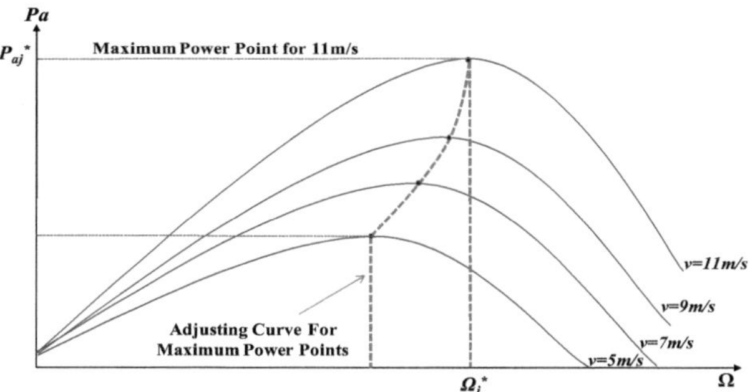

Fig.5. The shape of the aerodynamic power as a function of rotor speed for different values of wind speed. MPP highlighted

For a given wind speed, there is a single optimum rotor speed reference from which the maximum mechanical power can be extracted (see Fig.5). For this purpose, a sample of 20 relevant wind speed values v_j ($j = 1, \dots, 20$) has been selected a priori and the corresponding global maxima (Ω_j^*, P_{aj}^*) can be determined graphically [11]. In doing so, a set of 20 points $(\Omega_{rj}^*, P_{aj}^*)(j = 1, \dots, 20)$ was constructed. Then, a polynomial function of order n R(.), fitting in the least squares sense to the set of points, was constructed. $(\Omega_{rj}^*, P_{aj}^*)$ points, was constructed. For the wind turbine under consideration, characterised by the numerical parameters in Table 2 and the form of $C_P(\lambda)$presented in [11], the degree n=4 proved suitable for the data considered. The polynomial thus constructed is noted :

$$R(v) = w_4 v^4 + w_3 v^3 + w_2 v^2 + w_1 v + w_0 \quad \text{(I.55)}$$

Where the coefficients w_j have the numerical values given in Table 4. For the wind power system under consideration, the shape of R(v) is plotted in Fig.6 which will be referred to as the optimal wind speed power characteristic (OWRS).

Table.6. Numerical values of the coefficients of the polynomial $R(v)$.

Index	Value	Index	Value	Index	Value
w_0	45	w_2	9.996	w_4	−0.002
w_1	−0.04247	w_3	0.064		

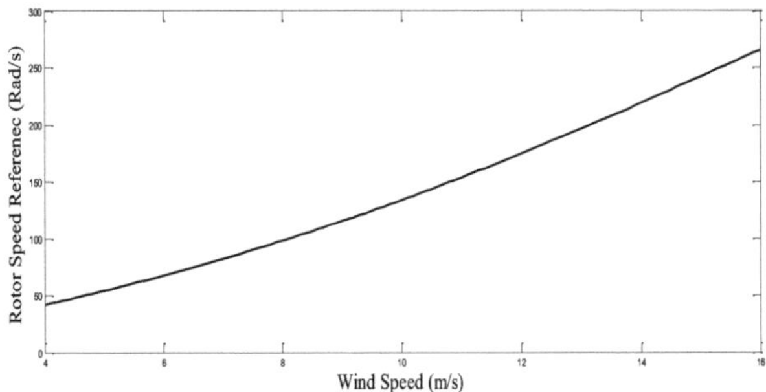

Fig.6. Optimum wind speed-rotor speed characteristic (OWRS).

B. Analysis of the rotor speed and flux controller design

We are interested in the problem of controlling the rotor speed and flux norm for the double-fed induction generator described by the model (36-43) which takes into account saturation and hysteresis of the magnetic characteristic. The speed and flux references (Ω_{ref},φ_{ref}) are bounded and derivable functions of time and their first two derivatives are available and bounded. These conditions can always be met by filtering the reference through second-order linear filters.

The controller will now be designed in two stages using the backstepping technique. First, let's introduce tracking errors:

$$e_1 = \Omega_{ref} - \Omega \quad (I.56)$$

$$z_1 = \varphi_{ref}^2 - \left(\varphi_{r\alpha}^2 + \varphi_{r\beta}^2\right) \quad (I.57)$$

Step 1: It follows from (36)-(43) that the errors e_1 and z_1 undergo the following differential equations:

$$\dot{e}_1 = \dot{\Omega}_{ref} - \frac{P}{J}\left(\varphi_{r\alpha} i_{s\beta} - \varphi_{r\beta} i_{s\alpha}\right) + \frac{T_L}{J} \quad (I.58)$$

$$\dot{z}_1 = 2\varphi_{ref}\dot{\varphi}_{ref} - 2\varphi_{r\alpha}\dot{\varphi}_{r\alpha} - 2\varphi_{r\beta}\dot{\varphi}_{r\beta} \quad (I.59)$$

Replacing (19)-(20) in (59) gives

$$\dot{z}_1 = 2\varphi_{ref}\dot{\varphi}_{ref} - 2\varphi_{r\alpha} V_{r\alpha} - 2\varphi_{r\beta} V_{r\beta} + 2R_r\left(\varphi_{r\alpha} i_{\mu\alpha} + \varphi_{r\beta} i_{\mu\beta}\right) - 2R_r\left(\varphi_{r\alpha} i_{s\alpha} + \varphi_{r\beta} i_{s\beta}\right) \quad (I.60)$$

In equations (58) the quantities $\mu_1 = \frac{P}{J}(\varphi_{r\alpha} i_{s\beta} - \varphi_{r\beta} i_{s\alpha})$stand as a virtual control signal. For the tracking error e_1 to cancel asymptotically, consider the following virtual control :

$$\mu_1 = c_1 e_1 + \dot{\Omega}_{ref} + \frac{T_L}{J} \tag{I.61}$$

Similarly, to control the tracking error$z_1$equation (60) suggests choosing the quantity $2\varphi_{r\alpha} V_{r\alpha} + 2\varphi_{r\beta} V_{r\beta}$ (considered as a virtual command) such that :

$$2\varphi_{r\alpha} V_{r\alpha} + 2\varphi_{r\beta} V_{r\beta} = d_1 z_1 + 2\varphi_{ref}\dot{\varphi}_{ref} + 2R_r(\varphi_{r\alpha} i_{\mu\alpha} + \varphi_{r\beta} i_{\mu\beta}) - 2R_r(\varphi_{r\alpha} i_{s\alpha} + \varphi_{r\beta} i_{s\beta}) \tag{I.62}$$

where c_1 and d_1 are positive real design parameters. Consider the Lyapunov candidate function defined by :

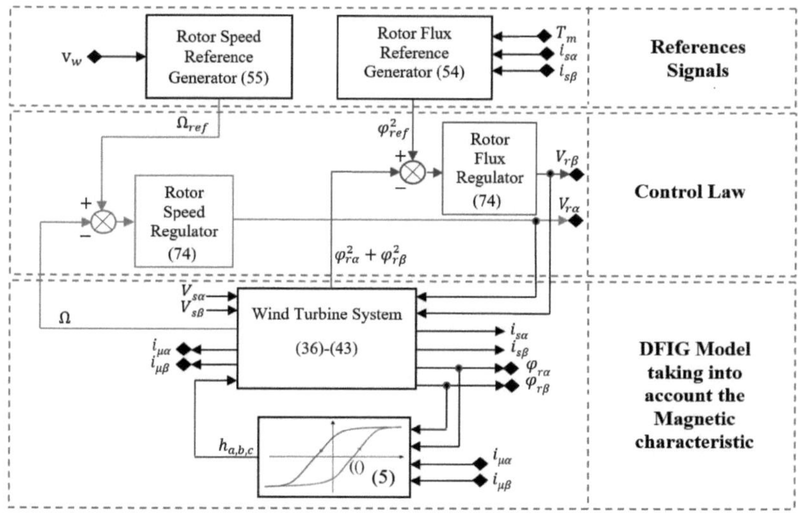

$$V_1 = \frac{1}{2}(e_1^2 + z_1^2) \tag{I.63}$$

Its time derivative is given by (using (61)-(62))

$$\dot{V}_1 = -c_1 e_1^2 - d_1 z_1^2 \tag{I.64}$$

In this step,μ_1 is taken as a first stabilising function, and the new tracking error denoted$e_2$is defined as:

$$e_2 = \mu_1 - \frac{P}{J}(\varphi_{r\alpha} i_{s\beta} - \varphi_{r\beta} i_{s\alpha}) \tag{I.65}$$

By replacing (61) in (65), the dynamics of the tracking error $e_1$is given by

$$\dot{e}_1 = -c_1 e_1 + e_2 \tag{I.66}$$

As a result, the time derivative of the Lyapunov candidate function (63) becomes smaller:

$$\dot{V}_1 = -c_1 e_1^2 - d_1 z_1^2 + e_1 e_2 \tag{I.67}$$

Step 2: The second step is to select the actual control signals, $V_{r\alpha}$ and $V_{r\beta}$ so that the error vector $[e_1, z_1, e_2]^T$ converge to zero. The time derivative of the tracking error e_2 is given by:

$$\dot{e}_2 = \dot{\mu}_1 - \frac{P}{J}\left(\dot{\varphi}_{r\alpha} i_{s\beta} + \varphi_{r\alpha} \dot{i}_{s\beta} - \dot{\varphi}_{r\beta} i_{s\alpha} - \varphi_{r\beta} \dot{i}_{s\alpha}\right) \tag{I.68}$$

Substituting (36)-(43) and (61) into (68) gives :

$$\dot{e}_2 = \mu_2 - \frac{P}{J}\left(V_{r\alpha} i_{s\beta} - \frac{1}{L_s} V_{r\beta} \varphi_{r\alpha} - V_{r\beta} i_{s\alpha} + \frac{1}{L_s} V_{r\alpha} \varphi_{r\beta}\right) \tag{I.69}$$

Where

$$\mu_2 = 2c_1(c_1 e_1 + e_2) + \ddot{\Omega}_{ref} + \frac{\dot{T}_L}{J} - \frac{P}{J}\Big(-R_r i_{\mu\alpha} i_{s\beta} + R_r i_{s\alpha} i_{s\beta} - \omega \varphi_{r\beta} i_{s\beta} - \frac{R_s}{L_s} \varphi_{r\alpha} i_{s\beta} + \frac{1}{L_s} V_{s\beta} \varphi_{r\alpha} + \frac{R_r}{L_s} i_{\mu\beta} \varphi_{r\alpha} - \frac{R_r}{L_s} i_{s\beta} \varphi_{r\alpha} - \frac{1}{L_s} \omega \varphi_{r\alpha}^2 + R_r i_{\mu\beta} i_{s\alpha} - R_r i_{s\beta} i_{s\alpha} - \omega \varphi_{r\alpha} i_{s\alpha} + \frac{R_s}{L_s} \varphi_{r\beta} i_{s\alpha} - \frac{1}{L_s} V_{s\alpha} \varphi_{r\beta} - \frac{R_r}{L_s} \varphi_{r\beta} i_{\mu\alpha} + \frac{R_r}{L_s} \varphi_{r\beta} i_{s\alpha} - \frac{1}{L_s} \omega_r \varphi_{r\beta}^2 - \frac{1}{L_s}\left(V_{s\alpha} \varphi_{r\beta} - V_{s\beta} \varphi_{r\alpha}\right)\Big) \tag{I.70}$$

To analyse the stability of the error system (e_1, z_1, e_2) consider the following extended Lyapunov candidate function :

$$V_2 = V_1 + \frac{1}{2} e_2^2 \tag{I.71}$$

Its time derivative along the trajectory of the state vector (e_1, z_1, e_2) is given by (using (67) and (69)) :

$$\dot{V}_2 = -c_1 e_1^2 - d_1 z_1^2 + e_2\left(e_1 + \mu_2 - \frac{P}{J}\left(V_{r\alpha}\left(i_{s\beta} + \frac{1}{L_s}\varphi_{r\beta}\right) - V_{r\beta}\left(i_{s\alpha} - \frac{1}{L_s}\varphi_{r\alpha}\right)\right)\right) \tag{I.72}$$

To ensure the overall and asymptotic stability of the error system, the control inputs $V_{r\alpha}, V_{r\beta}$ must be chosen as follows :

$$e_2\left[e_1 + \mu_2 - \frac{P}{J}\left(V_{r\alpha}\left(i_{s\beta} + \frac{1}{L_s}\varphi_{r\beta}\right) - V_{r\beta}\left(i_{s\alpha} - \frac{1}{L_s}\varphi_{r\alpha}\right)\right)\right] = -c_2 e_2^2 \tag{I.73}$$

Where c_2 is a positive design parameter. Then, using (62) and (73), we have :

$$\begin{bmatrix} V_{r\alpha} \\ V_{r\beta} \end{bmatrix} = \begin{bmatrix} \left(i_{s\beta} + \frac{1}{L_s}\varphi_{r\beta}\right) & -\left(i_{s\alpha} - \frac{1}{L_s}\varphi_{r\alpha}\right) \\ 2\varphi_{r\alpha} & 2\varphi_{r\beta} \end{bmatrix}^{-1}$$

$$\begin{bmatrix} \frac{J}{P}(c_2 e_2 + e_1 + \mu_2) \\ d_1 z_1^2 + 2\varphi_{ref}\dot{\varphi}_{ref} + 2R_r\left(\varphi_{r\alpha} i_{\mu\alpha} + \varphi_{r\beta} i_{\mu\beta}\right) - 2R_r\left(\varphi_{r\alpha} i_{s\alpha} + \varphi_{r\beta} i_{s\beta}\right) \end{bmatrix} \quad \text{(I.74)}$$

- **Theorem 1 (main result) :**

Consider the closed-loop system composed of the DFIG, described by the model (36)-(43), and the non-linear controller defined by the control law (74). The closed-loop error system described by (e_1, z_1, e_2) respectively given in (56, 57 and 65) is globally asymptotically stable with respect to the Lyapunov function (72). Consequently, all the errors disappear exponentially, whatever the initial conditions.

- **Proof of Theorem 1 :**

With the proposed control laws defined in (74), the time derivative of the Lyapunov function considered (71) is given by :$\dot{V}_2 = -c_1 e_1^2 - d_1 z_1^2 - c_2 e_2^2$ (I.75)
As $\dot{V}_2$ is a negative definite function of the state vector$(e_1, z_1, e_2)$the error system is globally asymptotically stable. This completes the proof of Theorem 1

I.1.5 Simulation results :

The simulation was carried out in Matlab/Simulink. The performance of the proposed new controller (the one that takes magnetic hysteresis into account) will be evaluated through several robustness tests. In the following, this controller will be referred to as HMC-OF. To demonstrate the supremacy of the HMC-OF, three comparisons will be carried out: the first test involves control strategies with a state-dependent optimal flux reference over control strategies with a nominal constant flux reference (see Table 2). The second test involves proving the performance of the HMC-OF against the standard controller assuming the DFIG magnetic characteristic is linear (in the following this controller will be referred to as LMC-NF), the latter working under a nominal flux reference. The final test will involve a comparison between the HMC-OF and the LMC-OF. Both controllers operate under an optimum flux reference, but the latter is based on a linear magnetic flux characteristic. The

DFIG under consideration is a 3MW whose characteristics are summarised in Table 4. The new controller design parameters are in Table 5, and the standard controller parameters are in Table 6.

Table 7: Machine electrical parameters.

Electric	**Index**	**Value**
Stator/rotor resistance	*Rs/ Rr*	0.455/0.62Ω
Leakage inductance S/Rotor	*Ls/ Lr*	0.0083/0.0081 H
Magnetising inductance	*Msr*	0.0078H
Inertia	*J*	0.3125kgm2
Viscous friction	*F*	6.73×10-1Nms-1

Table.8. New controller parameters.

Index	**Value**
C1	450
C2	240
d_1	5000

A. Simulation protocol :

The simulation protocol is set up to take into account a large variation in mean wind speed, as shown in Fig. 8. The algorithms used to calculate the flux and rotor speed references are discussed in subsections 4.1.2 and 4.1.1, respectively.

Numerous network faults will be applied to the HMC-OF in order to assess its robustness. To prove that it behaves correctly under real network operating conditions. These tests are divided into two categories

Parts: robustness to voltage dips and robustness despite frequency variation. The first fault considered is a three-phase voltage dip, which reduces the value of the main voltage by around 60% and lasts for 1s [4s-5s] as shown in Fig.9a. The second test was carried out considering a frequency variation [50-50.5 Hz]. This network fault is introduced at time 1s and lasts 0.5s as shown in Fig.9b.

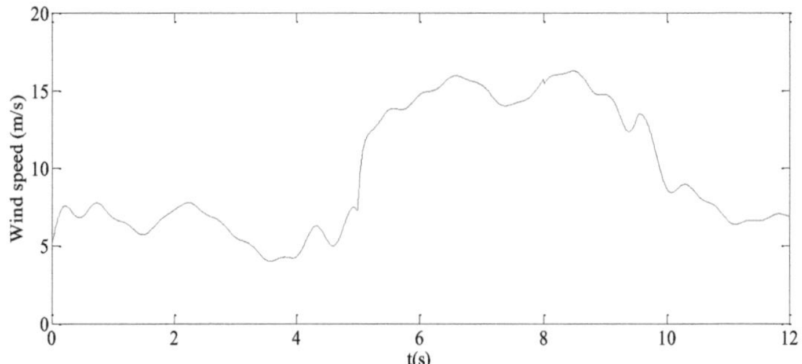

Fig.8. Wind speed profile considered

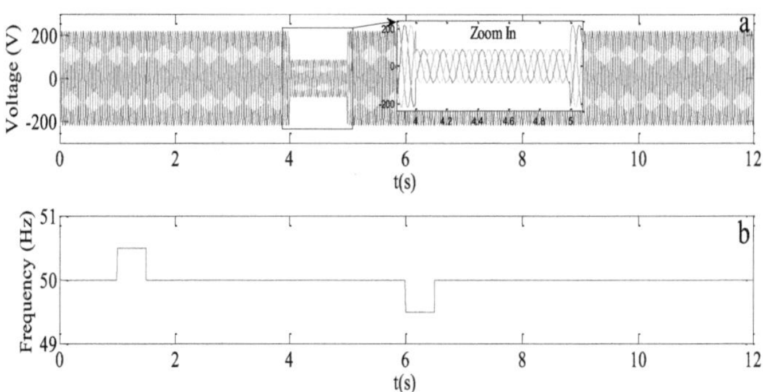

Fig.9a. Mains voltage at voltage dips.

Fig.9b. Network frequency profile.

B. Controller evaluation :

a) Supremacy of the HMC-OF strategy over the HMC-NF strategy (table 2)

The HMC-OF is implemented using equation (74). The corresponding design parameters are given by the numerical values in Table 3. Figure 10 shows that both the HMC-OF and the HMC-NF follow the given speed reference well, both are calculated using the optimal rotor speed curve (see Fig.3) ensuring the MPPT objective. However, the HMC-OF delivers more accuracy and exhibits less oscillation than the HMC-NF. Furthermore, Fig.10 highlights the good robustness of the proposed HMC-OF in the presence of the considered network faults (described in Figs.9a-9b), where it is clearly shown that the faults do not affect the tracking objectives.

Similarly, Fig.11 and Fig.12 show the HMC-OF and HMC-NF rotor flux tracking performances respectively. In fact, this figure illustrates the good behaviour of the proposed regulators even under the voltage dips and frequency deviation considered. Note that in Fig.11 the flux reference considered is variable as a function of the stator current in order to optimise stator joule losses (see subsection 4.1.1). In Fig.12, however, the flux reference considered is constant at around 1.1wb (equal to the nominal flux).

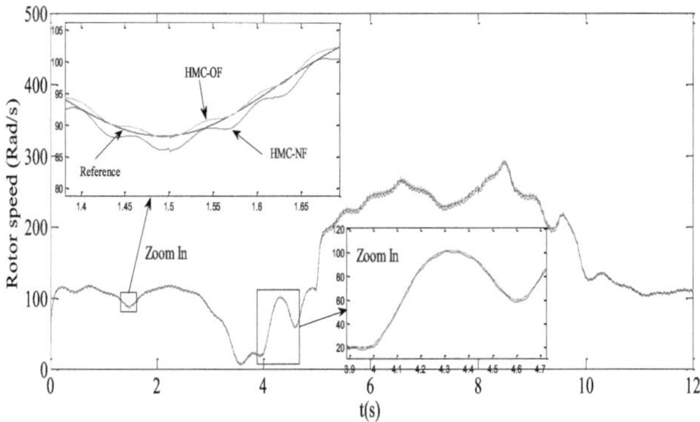

Fig.10. HMC-OF speed tracking performance. Solid: velocity reference. Dashed line: HMC-NF speed response. Dotted line: HMC-OF

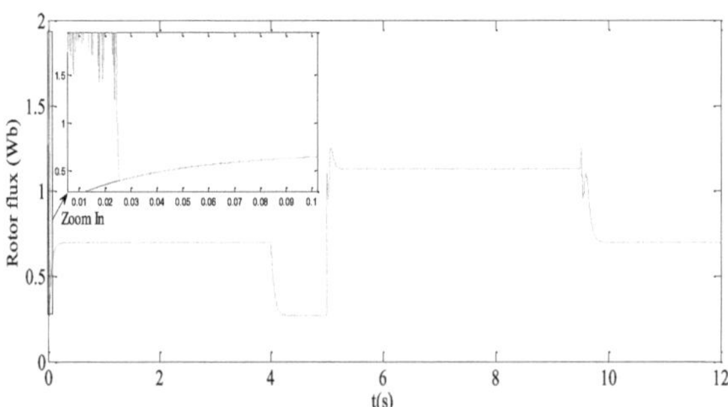

Fig.11. Rotor flux tracking performance. Solid line: Rotor flux reference. Dotted line: HMC-OF.

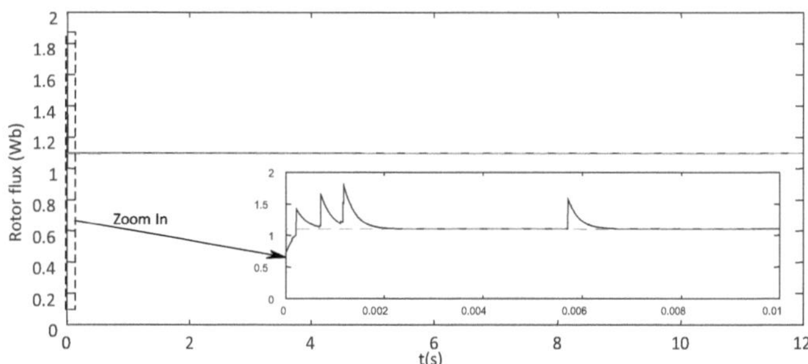

Fig.12. Rotor flux tracking performance. Solid line: Nominal rotor flux reference. Dotted line: HMC-NF.

Given that the difference between the HMC-OF and the HMC-NF is in the rotor flux, it is essential to carry out a power comparison to judge the advantage of one over the other. To do this, a power comparison is carried out between the power extracted from the generator working with the HMC-OF and the power extracted from the generator working with the HMC-NF; note that both power calculation systems take account of hysteresis/saturation phenomena. Figure 13 clearly shows that the power with the optimum flux reference (solid line) is greater than the power with the stable flux reference (dotted line), proving that with the optimum rotor flux, the joule power losses are lower and the power product is greater. Approximately 12% more power is collected using the optimum flux reference for the selected wind profile, attesting to the benefits of this controller.

The losses in joules are obtained by considering the expression$\frac{3}{2}R_sI_s^2$where I_s is the stator current standard. Fig.14 shows the evolution of these losses for the two controllers HMC-OF and HMC-NF, it is clearly shown that the HMC-OF implies lower losses than the HMC-NF, especially for low values of wind speed. This is the comparison between the HMC-OF and the HMC-NF.

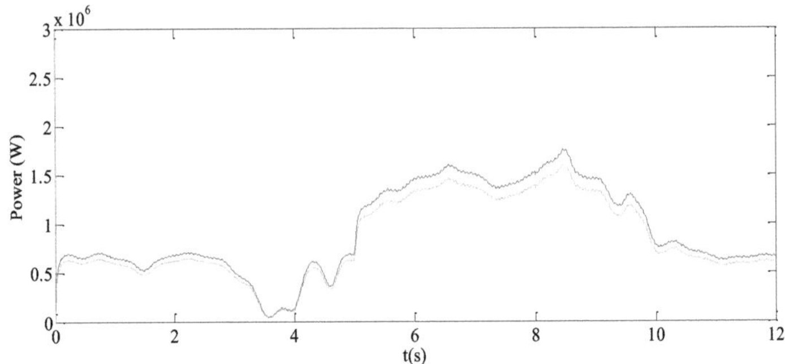

Fig. 13. Comparison of extracted power. Dotted line: HMC-NF. Solid line: HMC-OF.

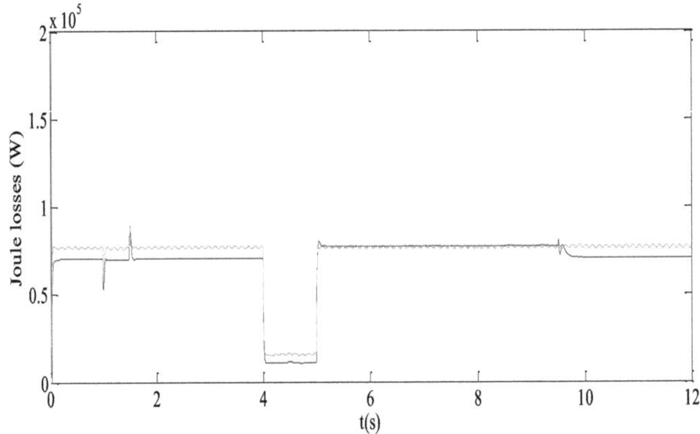

Fig.14. Joule losses. Dotted line: HMC-NF Solid line: HMC-OF.

b) Supremacy of the HMC-OF strategy over the LMC-NF strategy

In this subsection, the LMC-NF involving the nominal flow reference is considered for comparison purposes, the controller is obtained simply by considering in (31) :

- the function h constant and equal h=250, this is obtained by studying the linearisation of the hysteresis form in Fig.2.
- constant reference flux equal to 1.1Wb
- and other design parameters with the following values, which have proved practical for this controller :

C1=1200, C2=420, d1=100.

Fig. 15 shows the rotor speed tracking by the HMC-OF and the LMC-NF, clearly demonstrating the good response of the HMC-OF to the speed setpoint. On the other hand, the

LMC-NF is somewhat inaccurate at low rotor speeds; as soon as the speed increases, the tracking performance of the LMC-NF improves. Fig.16 shows that the LMC-NF tracks the rotor flux reference. However, small oscillations appear for different values of wind speed.

It should be noted that the LMC-NF simulation process working with the machine taking account of hysteresis/saturation phenomena required a great deal of effort and time to deliver such results, in other words no better results could be obtained with this configuration.

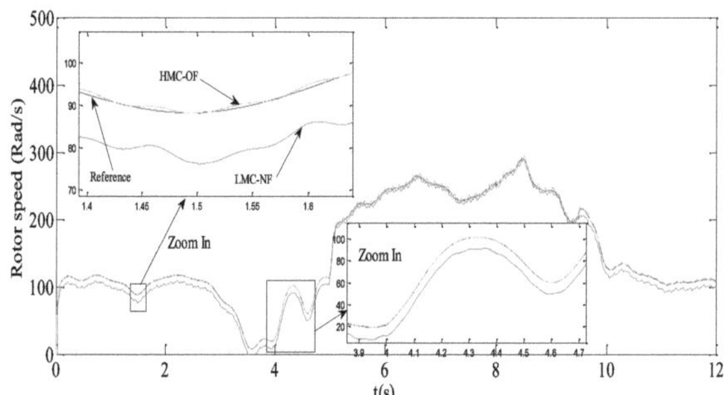

Fig.15. LMC-NF speed tracking performance. Solid: velocity reference. Dotted line: LMC-NF speed response. Dotted line: H MC-OF.

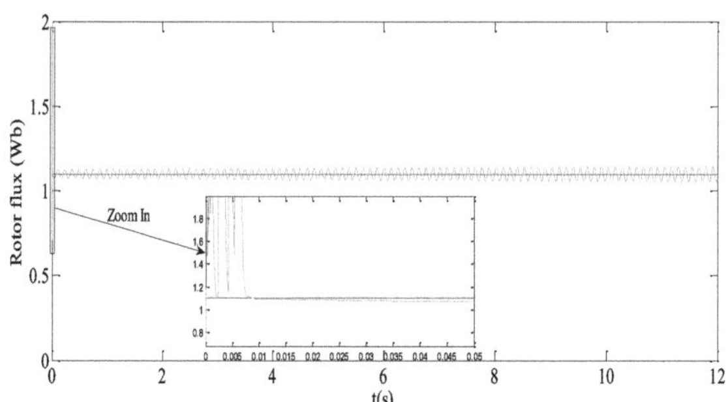

Fig.16. Rotor flux tracking performance. Solid line: Rotor flux reference. Dotted line LMC-NF

To complete the proof of the great interest of the HMC-OF, a power analysis has been carried out on Fig.17. The solid line shows the power delivered by the wind turbine operating with the HMC-OF, where the power is proven to be maximum since we are operating in MPPT mode. The dotted line shows the power delivered by the wind turbine driven by the LMC-NF.

We can see that there is a difference in power compared with the power delivered by the HMC-OF. This analysis proves that with the HMC-OF operating under real phenomena, more power can be extracted than with the standard controller, at both low and high wind speeds.
Fig. 18 clearly shows that HMC-OF involves lower joule losses than LMC-NF. This is the comparison between HMC-OF and LMC-NF.

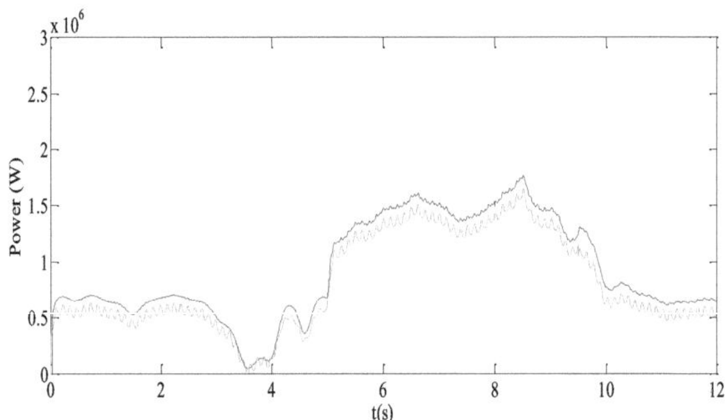

Fig.17. Comparison of extracted power. Dotted line: LMC-NF. Solid line: HMC-OF.

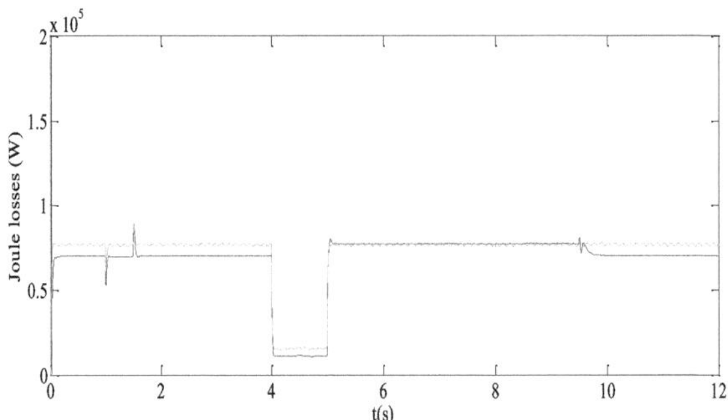

Fig.18. Joule losses. Dotted line: LMC-NF. Solid line: HMC-OF.

c) Supremacy of the HMC-OF strategy over the LMC-OF strategy

- LMC-OF is exactly the same shape as LMC-NF, except that :
- The flux reference is delivered by the optimal flux reference generator (see subsection 4.1.1).
- The regulator parameters are given the following values, which have proved to be practical: C_1 =450, C_2 =240, d_1 =100.

Figure 19 shows the rotor speed tracking performance of both the HMC-OF and the LMC-NF. It can clearly be seen that the HMC-OF gives greater accuracy, especially when operating at low wind speeds. Once the speed is increased, the tracking performance of the LMC-OF improves.

Figure 20 shows the rotor flux tracking performance of the LMC-OF. From this it can be seen that tracking is achieved around the nominal value of the rotor flux, and not with great accuracy elsewhere.

The power comparison between the HMC-OF and the LMC-OF shows that more power is extracted with the HMC-OF, particularly in low wind speed mode, as shown in Fig. 21.

Figure 22 clearly shows that HMC-OF involves lower joule losses than LMC-OF. This is the comparison between HMC-OF and LMC-OF.

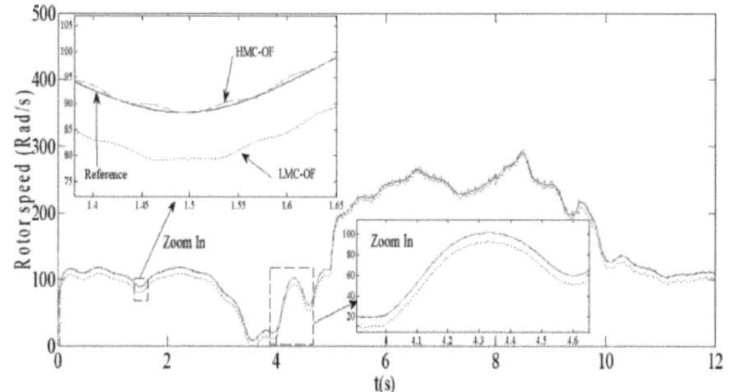

Fig.19. LMC-OF speed tracking performance. Solid: velocity reference. Dotted line: LMC-NF speed response. Dotted line: HMC-OF

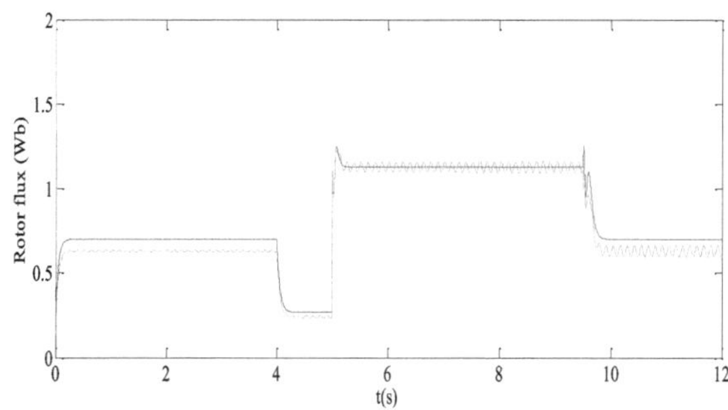

Fig.20. Rotor flux tracking performance. Solid line: Rotor flux reference. Dotted line LMC-OF

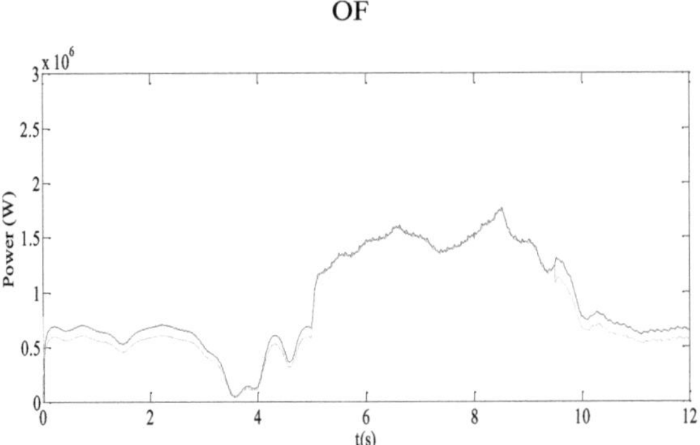

Fig.21. Comparison of extracted power. Dotted line: LMC-OF. Solid line: HMC-OF.

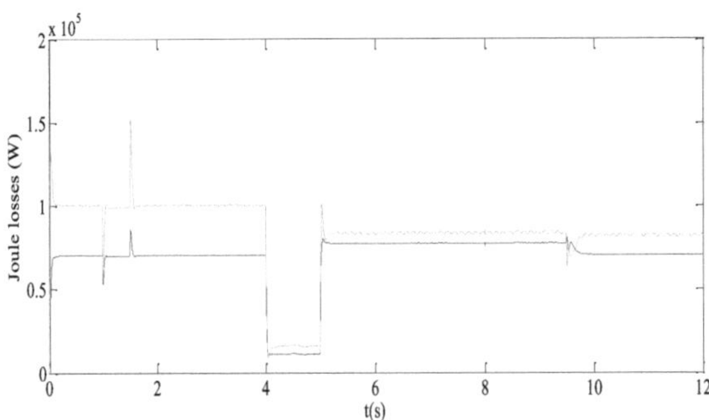

Fig.22. Joule losses. Dotted line: LMC-OF. Solid line: HMC-OF.

I.1.6 Conclusion of the chapter

In this chapter, we proposed a non-linear controller for the dual-fed induction generator operating in the presence of saturation and hysteresis of the magnetic characteristics. The aim of the control was to regulate the rotor speed and rotor flux. The latter are respectively delivered by an optimum rotor speed set-point generator, used to deliver a speed set-point corresponding to the maximum power available, and an optimum rotor flux set-point algorithm, designed to help this machine consume less power by Joule effect in the stator, and thereby generate more power. To this end, the HMC-OF described in (74) was designed using the backstepping technique, based on the model defined in (36-43). Theorem 1 formally establishes the global convergence of the errors $e_1 = \Omega_{ref} - \Omega$ and $z_1 = \varphi_{ref}^2 - (\varphi_{r\alpha}^2 + \varphi_{r\beta}^2)$ errors to zero. The controller has been analysed to determine sufficient conditions to properly track the given references. It is proved that the two errors are zero whatever the operating conditions. The overall stability of the system was studied and proved using Lyapunov theory. These theoretical results were confirmed by simulations on the Matlab/Simulink environment involving a wide range of wind speed variations. The robustness of the new controller was also assessed through several tests. This made it possible to check that the new controller maintains its good performance despite a variation in network frequency or undervoltage dips.

Chapter 1 references

[1] Ouadi, H., Giri, F., Elfadili, A., Dorleans, P., & Massieu, J. F; (2010). Induction Machine Control in Presence of Magnetic hysteresis Modelling and Speed reference tracking. *IFAC Proceedings Volumes*, *43*(10), 7-12.

[2] Coleman B. D., M. L. Hodgdon; (1987). On a class of constitutive relations for ferromagnetic hysteresis. *Arch. Rational Mech. Anal*, pp. 375-396.

[3] Du, J., Feng. Y., Su, C. Y., & Hu, Y. M; (2009). On the robust control of systems preceded by coleman-hodgdon hysteresis. *In IEEE 2009 Internatinal conference on control and automation* (pp 685-689). IEEE.

[4] Voros, J. (2009). Modeling and identification of hysteresis using special forms of the Coleman-Hodgdon model. *J. Electr. Eng,* 60(2), 100-105.

[5] Chen, X., Hisayama, T., &Su, C. Y. (2008, December). Adaptive control for continuous-time systems preceded by hysteresis. In *Decision and Control, 2008. CDC 2008. 47th IEEE Conference on* (pp. 1931-1936). IEEE.

[6] El Fadili, A., Giri, F., El Magri, A., Lajouad, R., Chaoui, F. (2013). Adaptive control strategy with flux reference optimization for sensorless induction motors. *Control Engineering Practice*, Vol. 26, May 2014, Pages 91-106.

[7] Akel, F., Ghennam, T., Berkouk, E. M., Laour, M. (2014). An improved sensorless decoupled power control scheme of grid connected variable speed wind turbine generator. *Energy Conversion and Management*, Vol. 78, Pages 584-594.

[8] Sarrias-Mena, Raúl, Fernández-Ramírez, L.M., García-Vázquez, C.A., Jurado, F. (2014). Fuzzy logic based power management strategy of a multi-MW doubly-fed induction generator wind turbine with battery and ultracapacitor. *Energy*, Vol 70, Pages 561-576.

[9] Dida, A., Benattous, D. (2015). Modeling and control of DFIG through back-to-back five levels converters based on neuro-fuzzy controller. *Journal of Control, Automation and Electrical Systems*. 26(5), 506-520.

[10] Hamane, B., Benghanem, M., Bouzid, M. A., Belabbes, A., Bouhamida, M., Draou, A. (2012). "Control for Variable Speed Wind Turbine Driving a Doubly Fed Induction Generator using Fuzzy-PI Control". *Energy Procedia,* 18 (2012) 476 - 485.

[11] Barra A, Ouadi H, Giri F. (2016). Sensorless Nonlinear Control of Wind Energy Systems With Doubly Fed Induction Generator. *Journal of control, Automation and Electrical Systems*. Vol 27. Issue 5. pages 562-578.

[12] Ouadi, H., Giri, F., Elfadili, A., &Dugard, L. (2010). Induction machine speed control with flux optimization. *Control Engineering Practice*, *18*(1), 55-66.

[13] El fadili, A., Giri, F., Ouadi, H., Dugard, L., & El Magri, A. (2009, August). Induction machine control in presence of magnetic saturation speed regulation with optimized flux reference. In *Control Conference (ECC), 2009 European* (pp. 2542-2547). IEEE.

[14] Macki, Jack W., Paolo Nistri, and Pietro Zecca (1993). Mathematical models for hysteresis. *SIAM review* 35.1: 94-123.

[15] Ouadi, H., Barra, A., & El Majdoub, K. (2017). Nonlinear control for grid connected wind energy system with multilevel inverter. *ARPN Journal of Engineering and Applied Sciences,* Vol. 12, N. 4.

Chapter II: Identifying the parameters of the asynchronous machine using the KALMAN filter

Summary of chapter II:

Induction machines are the most popular in the industrial environment because of their many advantages, including high power, robustness, ease of use and low cost. Their progress is largely due to the introduction of variable speed drives in the 1980s, which offered a wide variation in rotation frequency. In this chapter, we present a new method for estimating the internal parameters of the asynchronous cage machine. First of all, the constitutions of the asynchronous machine are mentioned, followed by a modelling of the machine. This was followed by a general overview of parameter estimation methods. Finally, the proposed method was applied. The results obtained are in good agreement with the parameters used in the simulation.

I.2.1. Introduction of the chapter

Induction machines are the most popular in the industrial environment because of their many advantages, including high power, robustness, ease of use and low cost. Their progress is largely due to the introduction of variable speed drives in the 1980s, which offered a wide variation in rotation frequency. In this chapter, we present a new method for estimating the internal parameters of the asynchronous cage machine. First of all, the constitutions of the asynchronous machine are described, followed by a modelling of the machine. This was followed by a general overview of parameter estimation methods. Finally, the proposed method was applied. The results obtained are in good agreement with the parameters used in the simulation.

I.2.2. Constitution of the asynchronous machine

An asynchronous machine as shown in Figalso known as an induction machine, consists of two main parts:

The stator: is the fixed part of the machine and consists of a cylindrical iron core with slots around its periphery. The slots contain the stator windings, which are generally three-phase windings. The stator windings are connected to a three-phase power supply, which produces a rotating magnetic field.

The rotor: is the rotating part of the machine and generally consists of a cylindrical iron core with slots around its periphery. The slots in the rotor contain conductor bars, usually made of copper or aluminium, which are short-circuited at both ends by end rings. These busbars are arranged in a specific pattern to form a squirrel-cage rotor.

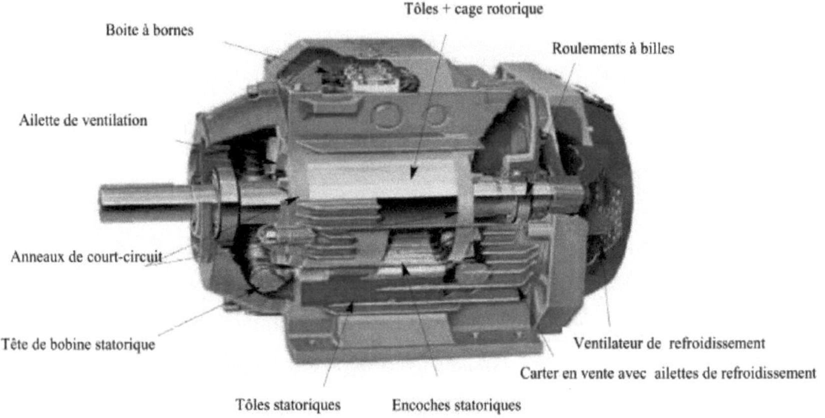

Fig .23. Composition of the asynchronous cage machine

When a three-phase voltage is applied to the stator windings, a rotating magnetic field is produced in the air gap between the stator and the rotor. The rotating magnetic field induces a current in the rotor bars, which produces a magnetic field that interacts with the magnetic field of the stator. The interaction between the two magnetic fields produces a torque on the rotor, causing it to rotate.

The asynchronous machine operates on the principle of electromagnetic induction, where a voltage is induced in the rotor conductors by the rotating magnetic field. The rotor is then subjected to a force that causes it to rotate, and the machine converts electrical energy into mechanical energy. The speed of the rotor is always slightly lower than the speed of the rotating magnetic field in the stator, which is known as slip. Slip is necessary to induce voltage in the rotor conductors and produce the torque required for rotation. The following section presents the different methods for estimating the parameters of these machines

I.2.3. Overview of methods for estimating asynchronous machine parameters

Induction machines are used in various applications such as conveyors and wind turbines. [[1]and their demand will increase with the massive shift towards electric vehicles. Despite the well-known qualities of these machines, during operation they are subject to electrical, magnetic, mechanical and thermal stresses, which cause changes to the motor's internal parameters. These changes, and their effects, are reflected in its amplitudes, mainly on flux, currents, speed and torque, which can be used to estimate the parameters of these machines.

In fact, having accurate values of the internal parameters of the IM will improve control performance. [2], [3] and will help monitor the health of the IM[4]. To achieve this, there are several methods for estimating parameters, which fall into two categories (See Fig).

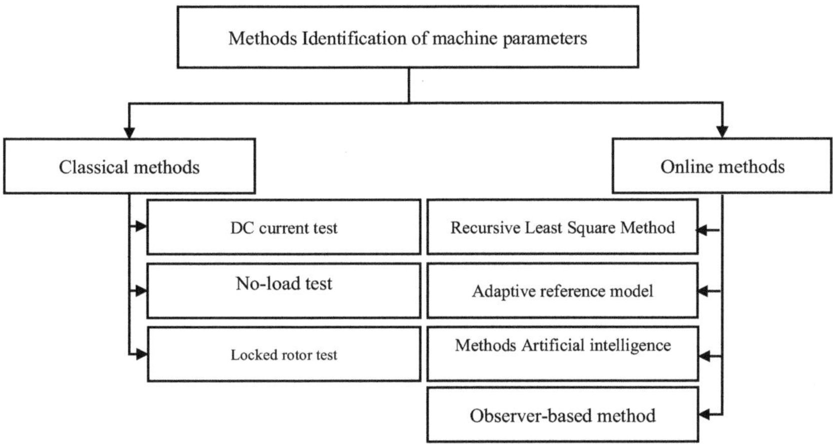

Fig.24.Methods for identifying the parameters of the asynchronous cage machine

A. Methods Classic s

The locked rotor test, the DC test and the no-load test are all used to determine the characteristics and parameters of an induction machine.[5]-[8].

i. The locked-rotor test consists of supplying the machine with a nominal voltage and frequency while the rotor is locked, in order to measure the stator impedance and the rotor resistance and reactance.

ii. The DC test involves applying a DC voltage to the stator winding and measuring the resulting current to determine the stator resistance.

iii. Finally, the no-load test involves running the machine at nominal voltage and frequency with the rotor unloaded, in order to measure the no-load current, input power and power factor. These measurements can be used to determine the machine's core losses and magnetising current.

These three tests are important for assessing the performance and efficiency of an induction machine. They are usually carried out during the manufacturing process or as part of routine maintenance.

B. Online methods

Because of the complexity of the conventional procedure, and the increase in winding temperature, the skin effect and flux saturation, which involves varying the parameters. It is therefore necessary to identify them in real time. Numerous attempts have been proposed in the literature, where various techniques are applied. In this context, algorithms based on the mathematical model of the engine or on artificial intelligence are discussed in the following.

C. Estimation method based on the RCM algorithm

Among the various identification algorithms, the recursive least squares (RLS) algorithm is widely used because it is fast, efficient and easy to implement. However, the RLS estimation algorithm requires the following inputs: rotational speed, stator currents and voltages, and their derivatives. Due to the noise and harmonics of these inputs, filters are required. The detailed procedure for implementing the MCR algorithm is shown in Fig.

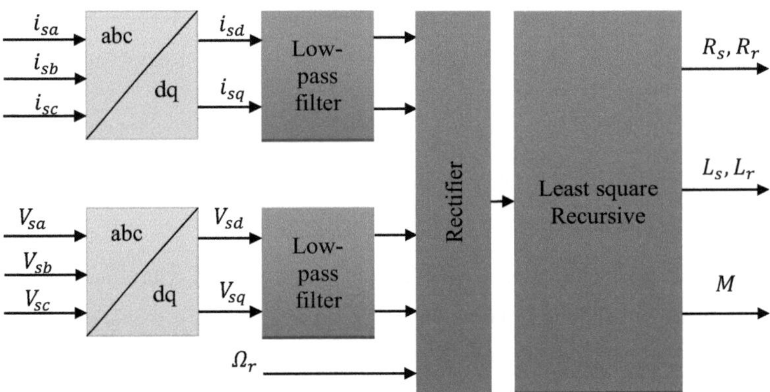

Fig .25.Implementation of the MCR algorithm for parameter identification

In addition, the forgetting factor will largely affect the performance of the MCR algorithm. If the forgetting factor decreases, the data will largely affect the results. Thus, the identified parameters converge rapidly, but stability is reduced, i.e. the algorithm is likely to diverge. On the contrary, if the forgetting factor increases, the convergence process is slower, and consequently the identification takes longer to follow the real values, while stability is improved. In particular, when the forgetting factor is fixed at 1, the estimation algorithm degenerates into a conventional RCM. Therefore, an appropriate forgetting factor needs to be defined, taking into account the trade-off between convergence rate and algorithm stability. In general, the forgetting factor can be set between 0.9 and 1.0.

D. Method based on the adaptive system with reference model

The Model-Referenced Adaptive System (MRAS) technique is a relatively mature method that has been applied to parameter identification because of its simple structure and ease of implementation. The basic principle of MRAS is described in Fig. In this technique, a real engine is included as a reference model, while the mathematical model of the engine that contains the unknown parameters is taken as the adjustable model. The inputs to both models are the same. Using an adaptive law, the estimated parameters will converge to their true values by reducing the error vector between the outputs of the reference and adaptive models to zero. Consequently, the unknown parameters can be identified online by applying the MRAS law.

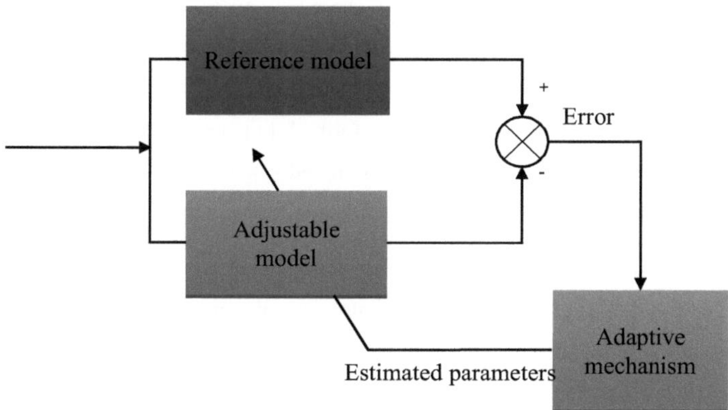

Fig .26.Principle of the MRAS method for identifying induction machine parameters

Based on the formulation of the error signal in the MRAS model, various identification methods based on the MRAS technique have been developed. Methods based on electromagnetic torque, active power, reactive power [9]-[11],rotor flux [1[12]stator voltage [[10], [113]the electromagnetic field [1[12]etc.

E. Observer-based methods

An observer can also be used to determine the motor parameters. Typically, the actual motor measurements are subtracted from the observer outputs before being multiplied by a matrix L. The multiplication is then added to the observer's equations of state, creating a so-called LUENBERGER observer, as follows

$$\begin{cases} \dot{\hat{x}}(t) = A\hat{x}(t) + L(\hat{y}(t) - y(t)) + Bu(t) \\ y(t) = C\hat{x}(t) \end{cases} \tag{II.1}$$

Where u(t) represents the inputs at time t, x(t) represents the state variables, y(t) represents its outputs, and $\hat{x}(t)$ represents the estimated variables. As the motor model is non-linear, it is then approximated using Taylor expansion to linearise the model around the previous estimated state for machine parameter identification. The linearisation must be a good approximation of the non-linear model to ensure the validity of this approach. In [[14], [115]where the unknown parameters have been augmented into state variables, extended LUENBERGER observers have been proposed to identify the motor parameters. Consequently, the size of the state variables has been increased.

In addition, the extended Kalman filter (FKE) is an optimisation algorithm that can overcome the noise sensitivity problems of the MCR algorithm. For example, the ELO and FKE observers attempt to solve the non-linearity problem using a linear approximation, where linearisation of the current state estimate is performed. In general, the FKE algorithm includes prediction and filtering steps, where future predicted states are determined in the prediction step and estimated states are obtained by adding corrections to the predicted states.

In [[16]the author presents the FKE algorithm, which is a method for calculating the inverse rotor time constant from recorded stator currents and rotor speed. In this case, the relationship between magnetising current and inductance was used to create a look-up table to determine the magnetising inductance. Furthermore, in [[17]the FKE and MCR algorithms were combined to determine the motor parameters, with the former (i.e. the FKE algorithm) focusing more on the rotor resistance and the latter (i.e. the MCR algorithm) on estimating the rotor inertia constant, damping constant and disturbed load torque. However, despite the various advantages of FKE, its implementation requires a significant computational load.

F. Methods based on artificial intelligence

There are several AI-based methods that can be used for estimating induction machine parameters, including:

Artificial Neural Networks (ANNs): These are widely used for estimating induction machine parameters because of their ability to learn complex relationships between inputs and outputs. ANNs can be trained using historical data and used to predict machine parameters [18]-[220].

Fuzzy logic: Fuzzy logic is a mathematical framework that can be used to model uncertainty and imprecision. Fuzzy logic can be used with other AI-based methods to estimate the parameters of induction machines[221], [222].

Genetic algorithms (GAs): GAs are optimisation algorithms that can be used to search for the optimal set of parameters. GAs are often used in conjunction with other AI-based methods to estimate the parameters of induction machines. [223].

Particle Swarm Optimisation (PSO): PSO is another optimisation algorithm that can be used to estimate the parameters of induction machines. PSO is based on the motion of particles in a search space and can be used to find the optimal set of parameters[224], [225].

These AI-based methods can be used individually or in combination with each other to estimate induction machine parameters. The choice of method(s) depends on the specific application and the available data.

I.2.4. Modelling the asynchronous machine

Electrical machines are very complex systems. That it is not possible to find a suitable model that captures all the basic phenomena to describe the behaviour of these machines; therefore, the following simplifying assumptions are used to find a mathematical model. [226]:

i) The air gap is of uniform thickness, and the notch effect is negligible.

ii) Magnetic circuit saturation, hysteresis and eddy currents are negligible.

iii) The winding resistances do not vary with temperature and the skin effect is negligible.

iv) It is recognised that the magnetomotive forces created by each phase of the two frames are sinusoidally distributed.

Where the equations of the machine in the three phases are given by :

$$V_{sabc} = R_s I_{sabc} + \dot{\varphi}_{sabc} \quad \text{(II.2)}$$

$$V_{rabc} = R_r I_{rabc} + \dot{\varphi}_{rabc} \quad \text{(II.3)}$$

The vectors $V_{sabc}, V_{rabc}, I_{sabc}, I_{rabc}, \dot{\varphi}_{sabc}, \dot{\varphi}_{rabc}$ are defined as

$$V_{sabc} = \begin{bmatrix} v_{sa} \\ v_{sb} \\ v_{sc} \end{bmatrix} \quad I_{sabc} = \begin{bmatrix} i_{sa} \\ i_{sb} \\ i_{sc} \end{bmatrix} \quad \varphi_{sabc} = \begin{bmatrix} \varphi_{sa} \\ \varphi_{sb} \\ \varphi_{sc} \end{bmatrix}$$
$$V_{rabc} = \begin{bmatrix} v_{ra} \\ v_{rb} \\ v_{rc} \end{bmatrix} \quad I_{rabc} = \begin{bmatrix} i_{ra} \\ i_{rb} \\ i_{rc} \end{bmatrix} \quad \varphi_{rabc} = \begin{bmatrix} \varphi_{ra} \\ \varphi_{rb} \\ \varphi_{rc} \end{bmatrix} \quad \text{(II.4)}$$

(v_{sa}, v_{sb}, v_{sc}) Stator voltages.

(v_{ra}, v_{rb}, v_{rc}) Rotor voltages.

(i_{sa}, i_{sb}, i_{sc}) Stator currents.

(i_{ra}, i_{rb}, i_{rc}) Rotor currents.

$(\varphi_{sa}, \varphi_{sb}, \varphi_{sc})$ Stator fluxes.

$(\varphi_{ra}, \varphi_{rb}, \varphi_{rc})$ Rotor flux.

The total flux of the induction motor is related to the currents by the following equations:

$$\varphi_{sabc} = L_{ss} I_{sabc} + M_{sr} I_{rabc} \quad \text{(II.5)}$$

$$\varphi_{rabc} = L_{rr} I_{rabc} + M_{rs} I_{sabc} \text{(II.6)}$$

With

$$L_{ss} = \begin{bmatrix} L_s & M_s & M_s \\ M_s & L_s & M_s \\ M_s & M_s & L_s \end{bmatrix}$$

$$L_{rr} = \begin{bmatrix} L_r & M_r & M_r \\ M_r & L_r & M_r \\ M_r & M_r & L_r \end{bmatrix}$$

$$M_{sr} = M_{rs}^T = M_{sr} \begin{bmatrix} cos(\theta) & cos\left(\theta - \frac{4\pi}{3}\right) & cos\left(\theta - \frac{2\pi}{3}\right) \\ cos\left(\theta - \frac{2\pi}{3}\right) & cos(\theta) & cos\left(\theta - \frac{4\pi}{3}\right) \\ cos\left(\theta - \frac{4\pi}{3}\right) & cos\left(\theta - \frac{2\pi}{3}\right) & cos(\theta) \end{bmatrix}$$

Hence

M_s Is the inductance between two stator phases

M_rIs the inductance between two rotor phases

L_s Is the stator inductance

L_r Is the rotor inductance

As the model given by the three phases is complicated, it is necessary to use a transformation of the system. Consequently, the equations to be derived are based on the transformation of the mathematical model of the three-phase model into a two-phase model using the transformation techniques of Clarke and Park.

A. Electrical equations and mechanical equations of the asynchronous machine

The AM electrical equations are based on the transformation of the mathematical model from three-phase to two-phase using the Park transformation technique. Where the motor equations with rotor pulsation $\omega_r = \omega_s - p\Omega_r$ in both phases are given by [226], [227]:

- Equations of the stator and rotor voltages in the dq reference frame

$$V_{sd} = R_s i_{sd} + \dot{\varphi}_{sd} - \omega_s \varphi_{sq} \quad \text{(II.7)}$$

$$V_{sq} = R_s i_{sq} + \dot{\varphi}_{sq} + \omega_s \varphi_{sd} \quad \text{(II.8)}$$

$$V_{rd} = R_r i_{rd} + \dot{\varphi}_{rd} - \omega_r \varphi_{rq} \quad \text{(II.9)}$$

$$V_{rq} = R_r i_{rq} + \dot{\varphi}_{rq} + \omega_r \varphi_{rd} \quad \text{(II.10)}$$

- Equations of the stator and rotor fluxes in the dq reference frame

$$\varphi_{sd} = L_s i_{sd} + M i_{rd} \quad \text{(II.11)}$$

$$\varphi_{sq} = L_s i_{sq} + M i_{rq} \quad \text{(II.12)}$$

$$\varphi_{rd} = L_s i_{rd} + M i_{sd} \quad \text{(II.13)}$$

$$\varphi_{rq} = L_s i_{rq} + M i_{sq} \quad \text{(II.14)}$$

The electromagnetic torque is obtained from the following equations:

$$\begin{aligned} C_e &= pM(\varphi_{sd} i_{sq} - \varphi_{sq} i_{sd}) \\ C_e &= pM(\varphi_{rq} i_{rd} - \varphi_{rd} i_{rq}) \\ C_e &= \frac{pM}{L_r}(\varphi_{rd} i_{sq} - \varphi_{rq} i_{sd}) \\ C_e &= \frac{pM}{L_r}(\varphi_{sq} i_{rd} - \varphi_{sd} i_{rq}) \end{aligned} \quad \text{(II.15)}$$

Mechanical speed is deduced from the fundamental law of general mechanics as follows:

$$J_m \dot{\Omega}_r = C_e - f_m \Omega_r - C_r \quad \text{(II.16)}$$

Where

C_e Electromagnetic torque

C_r Resistive torque

f_m Fiction coefficient

J_m Moment of inertia

I.2.5. Application

IM parameters are traditionally estimated using three well-known offline methods. The DC test, the no-load test and the locked-rotor test [223]. These methods are simple, but very approximate. Moreover, as the machine is coupled to a mechanical load, carrying out these tests in an industrial environment is quite difficult. In addition, the service must be interrupted while these tests are carried out [228]. To overcome these challenges, a wide range of techniques for determining induction motor parameters can be found in the literature [2[29], [330].

Because it is fast, efficient and easy to implement, the recursive least squares (RLS) algorithm is a popular identification technique. Several studies of induction machine parameter estimation based on the RLS algorithm have been developed in this field. In [3[31]the RCM is used to estimate the steady-state parameters of the IM. In [3[32]a BUTTERWORTH filter is used to reduce the effect of noise and improve the estimation. In [333]-[335]the MCR algorithm is used with a FOC controller to estimate time-varying MI parameters and improve control performance. These methods give good results. However, the MCR algorithm is known to be sensitive to noise, and these studies are carried out in the steady state of the machine when the speed is considered to be constant.

This work is devoted to estimating the internal parameters of the IM, which include the stator resistance $R_s$rotor resistance $R_r$stator induction Ls, rotor induction $L_r$and the magnetising inductance$M$, . Compared to previous works, the present study aims at estimating the parameters of the MI while the speed is variable using the linear KALMAN filter considering a variable speed profile. In the following, we are interested in modelling the IM in the αβ reference frame while considering the various simplifying assumptions. This model is used to develop a new regression equation for parameter estimation. Then, the KF used for parameter estimation is introduced. The results of the proposed method are then discussed. Finally, a general conclusion closes this chapter.

A. Asynchronous machine model used for parameter estimation

The first step in the estimation process is to define an appropriate model of the system. This model should reflect as accurately as possible all the phenomena that the designer is seeking to highlight, in order to predict the behaviour of the physical system in the dynamic and static regimes. However, electrical machines are too complex to be able to include all the physical phenomena they undergo in the modelling. It is therefore essential to introduce a few classic simplifying hypotheses that in no way alter the validity of the machine model.

The model is represented in a dq reference frame with equations IV-7 to IV-15 is used for linear control such as direct and indirect flux orientation control which requires the stator angle. However, for non-linear control and state and parameter estimation, the most commonly used model is the αβ model which does not require stator angle estimation. To construct the αβ model just from the dq model, the stator angle must be set to zero $\theta_s$and therefore $\omega_s = 0$, hence the rotor pulsation $\omega_r = p\Omega_r$. Therefore, the model of the squirrel cage induction machine is given by the following equations. [3[36]:

$$\frac{d\varphi_{s\alpha}}{dt} = V_{s\alpha} - R_s i_{s\alpha} \tag{II.17}$$

$$\frac{d\varphi_{s\beta}}{dt} = V_{s\beta} - R_s i_{s\beta} \tag{II.18}$$

$$V_{r\alpha} = 0 = R_r i_{r\alpha} - \dot{\omega}_r \varphi_{r\beta} + \frac{d\varphi_{r\alpha}}{dt} \tag{II.19}$$

$$V_{r\beta} = 0 = R_r i_{r\beta} + \dot{\omega}_r \varphi_{r\alpha} + \frac{d\varphi_{r\beta}}{dt} \tag{II.20}$$

$$\varphi_{s\alpha} = L_s i_{s\alpha} + M i_{r\alpha} \tag{II.21}$$

$$\varphi_{s\beta} = L_s i_{s\beta} + M i_{r\beta} \tag{II.22}$$

$$\varphi_{r\alpha} = M i_{s\alpha} + L_r i_{r\alpha} \tag{II.23}$$

$$\varphi_{r\beta} = M i_{s\beta} + L_r i_{r\beta} \tag{II.24}$$

$(\varphi_{s\alpha}\varphi_{s\beta})$ Stator flux in the reference frame $\alpha\beta$.

$(\varphi_{r\alpha}, \varphi_{r\beta})$ Rotor flux in the reference frame $\alpha\beta$.

$(i_{s\alpha}, i_{s\beta})$ Stator currents in the reference frame $\alpha\beta$.

$(i_{r\alpha}, i_{r\beta})$ Rotor currents in the $\alpha\beta$.

ω_r Rotor pulse.

According to (IV-21) and (IV-22), we have

$$i_{r\alpha} = \frac{\varphi_{s\alpha} - L_s i_{s\alpha}}{M} \quad \text{(II.25)}$$

$$i_{r\beta} = \frac{\varphi_{s\beta} - L_r i_{s\beta}}{M} \quad \text{(II.26)}$$

By replacing (IV-25) in (IV-23) and (IV-26) in (IV-24), we obtain.

$$\varphi_{r\alpha} = \frac{L_r}{M}\varphi_{s\alpha} + \left(M - \frac{L_r L_s}{M}\right) i_{s\alpha} \quad \text{(II.27)}$$

$$\varphi_{r\beta} = \frac{L_r}{M}\varphi_{s\beta} + \left(M - \frac{L_r L_s}{M}\right) i_{s\beta} \quad \text{(II.28)}$$

Replacing (IV-27) and (IV-28) in (IV-19) :

$$\frac{di_{s\alpha}}{dt} = -\frac{R_r L_s + R_s L_r}{\sigma L_s L_r} i_{s\alpha} - p\Omega_r i_{s\beta} + \frac{R_r}{\sigma L_s L_r}\varphi_{s\alpha} + \frac{p\Omega_r}{\sigma L_s}\varphi_{s\beta} + \frac{V_{s\alpha}}{\sigma L_s} \quad \text{(II.29)}$$

Replacing (IV-27) and (IV-27) in (IV-20) :

$$\frac{di_{s\beta}}{dt} = p\Omega_r i_{s\alpha} - \frac{R_r L_s + R_s L_r}{\sigma L_s L_r} i_{s\beta} - \frac{p\Omega_r}{\sigma L_s}\varphi_{s\alpha} + \frac{R_r}{\sigma L_s L_r}\varphi_{s\beta} + \frac{V_{s\beta}}{\sigma L_s} \quad \text{(II.30)}$$

With

$$\sigma = \left(1 - \frac{M^2}{L_s L_r}\right)$$

$$\omega_r = p\Omega_r$$

$\Omega_r$is the rotor speed, and σ is the leakage factor.

By integration of (IV-17) and (IV-18), we have

$$\varphi_{s\alpha} = -R_s \int_0^T i_{s\alpha} dT + \int_0^T V_{s\alpha} dt + C_1 \quad \text{(II.31)}$$

$$\varphi_{s\beta} = -R_s \int_0^T i_{s\beta} dT + \int_0^T V_{s\beta} dT + C_2 \quad \text{(II.32)}$$

And using (IV-31) and (IV-23) in (IV-29) :

$$\frac{di_{s\alpha}}{dT} = -\frac{R_r L_s + R_s L_r}{\sigma L_s L_r} i_{s\alpha} - p\Omega_r i_{s\beta} - \frac{R_r R_s}{\sigma L_s L_r}\int_0^T i_{s\alpha} dT + \frac{R_r}{\sigma L_s L_r}\int_0^T V_{s\alpha} dt - \frac{pR_s}{\sigma L_s}\Omega_r \int_0^T i_{s\beta} dt +$$
$$\frac{p}{\sigma L_s}\Omega_r \int_0^T V_{s\beta} dt + \frac{1}{\sigma L_s} V_{s\alpha} + \frac{pC_2}{\sigma L_s}\Omega_r + C_3 \quad \text{(II.33)}$$

By integrating equation (IV-33), we obtain

$$i_{s\alpha} = -\frac{R_r L_s + R_s L_r}{\sigma L_s L_r}\int_0^T i_{s\alpha}dt - p\int_0^T \Omega_r i_{s\beta}dt - \frac{R_r R_s}{\sigma L_s L_r}\int_0^T\int_0^T i_{s\alpha}dt^2 + \frac{R_r}{\sigma L_s L_r}\int_0^T\int_0^T V_{s\alpha}dt^2$$
$$-\frac{pR_s}{\sigma L_s}\int_0^T\left(\Omega_r\int_0^T i_{s\beta}dt\right)dt + \frac{p}{\sigma L_s}\int_0^T\left(\Omega_r\int_0^T V_{s\beta}dt\right)dt + \frac{1}{\sigma L_s}\int_0^T V_{s\alpha}dt$$
$$+\frac{pC_2}{\sigma L_s}\int_0^T \Omega_r dt + C_3\cdot t + C_4$$

(II.34)

Equation (IV-34) does not use the rotor flux signals, which are considered difficult to measure, and is linear in the coefficients θ and can be expressed as equation (IV-35). The cladding form of this machine model allows the FK identification procedure discussed in the next section to calculate the electrical parameters of the motors.

$$Y = \Psi\cdot\theta \qquad \text{(II.35)}$$

With

$$\psi = [\psi_1 \quad \psi_2 \quad \psi_3 \quad \psi_4 \quad \psi_5 \quad \psi_6 \quad \psi_7 \quad \psi_9 \quad \psi_9 \quad \psi_{10}]$$
$$\theta = [-\theta_1 \quad -\theta_2 \quad -\theta_3 \quad \theta_4 \quad -\theta_5 \quad \theta_6 \quad \theta_7 \quad \theta_9 \quad \theta_9 \quad \theta_{10}]^T$$
$$Y = i_{s\alpha}$$

Hence

$\theta_1 = \left(\frac{R_s}{\sigma L_s} + \frac{R_r}{\sigma L_r}\right)$	$\theta_2 = p$	$\theta_3 = \frac{R_s R_r}{\sigma L_s L_r}$	$\theta_4 = \frac{R_r}{\sigma L_s L_r}$
$\theta_5 = \frac{pR_s}{\sigma L_s}$	$\theta_6 = \frac{p}{\sigma L_s}$	$\theta_7 = \frac{1}{\sigma L_s}$	$\theta_8 = \frac{pC_2}{\sigma L_s}$
$\theta_9 = C_3$	$\theta_{10} = C_4$		

And

$\psi_1 = \int_0^T i_{s\alpha}dt$	$\psi_2 = \int_0^T \Omega_r i_{s\beta}dt$	$\psi_3 = \int_0^T\int_0^T i_{s\alpha}dt^2$	$\psi_4 = \int_0^T\int_0^T V_{s\alpha}dt^2$

$\psi_5 = \int_0^T \left(\Omega_r \int_0^T i_{s\beta} dt\right) dt$	$\psi_6 = \int_0^T \left(\Omega_r \int_0^T V_{s\beta} dt\right) dt$	$\psi_7 = \int_0^T V_{s\alpha} dt$	$\psi_8 = \int_0^T \Omega_r dt$
$\psi_9 = t$	$\psi_{10} = 1$		

Finally, we can obtain the formulae for the motor parameters, which include the stator and rotor resistances $\hat{R}_s$ and $\hat{R}_r$ the magnetic inductances $\hat{L}_s, \hat{L}_r$ and $\hat{M}$ and the magnetic leakage coefficient $\hat{\sigma}$ assuming $L_s = L_r$ And using the relationships below.

$\hat{p} = \hat{\theta}_2$	$\hat{R}_s = \dfrac{\hat{\theta}_5}{\hat{\theta}_6}$	$\hat{R}_r = \dfrac{\hat{\theta}_1}{\hat{\theta}_7} - \dfrac{\hat{\theta}_5}{\hat{\theta}_6}$
$\hat{L}_s = \hat{L}_r = \hat{L}_{sr} = \dfrac{\hat{\theta}_1\hat{\theta}_6 - \hat{\theta}_5\hat{\theta}_7}{\hat{\theta}_4\hat{\theta}_6}$	$\hat{\sigma} = \dfrac{\hat{\theta}_4\hat{\theta}_6\hat{\theta}_7}{\hat{\theta}_1\hat{\theta}_6 - \hat{\theta}_5\hat{\theta}_7}$	$M^2 = (1-\sigma)\hat{L}_{sr}^2$

B. KALMAN filter for parameter estimation

In this section, we describe the KALMAN filter algorithm used to estimate the parameters of the asynchronous machine. We consider a stochastic system defined by :

$$y(k) = \psi^T(k)\theta + v(k) \tag{II.36}$$

Where $y(k)$ is the observation, $v(k)$are the noise signals, $\psi(k)$ is the regressor, θ are the system parameters.

To estimate the unknown parameters $\hat{\theta}$we introduce the following recursive KALMAN filter [337], [338].

$$\hat{\theta}(k) = \hat{\theta}(k-1) + \frac{P(k-1)\psi^T(k)}{R+\psi(k)P(k-1)\psi^T(k)}\left(y(k) - \psi(k)\hat{\theta}(k-1)\right) \tag{II.37}$$

$$P(k) = P(k-1) - \frac{P(k-1)\psi_k^T\psi(k)P(k-1)}{R+\psi(k)P(k-1)\psi^T(k)} + Q \tag{II.38}$$

Or,

- $Q = Ew(k)w^T(k)$is obtained by considering the following model of the system parameters $\theta(k) = \theta(k-1) + w(k)$
- $R = Ev^2(k)$

Where Q > 0, R > 0, $\hat{\theta}(0)$and $P(0)$ are deterministic and can be chosen arbitrarily. According to [3[38]we can estimate the time-varying parameters by estimating Q and R. However, in this work we assume that the parameters are constant.

Note that if Q = 0 and R = 1, we obtain the recursive least squares algorithm.

I.2.6. Simulation and results

The simulation of the proposed estimation method is performed with MATLAB using the numerical values of the parameters as follows:

Table 9: Nominal values of asynchronous machine parameters

Parameter	Nominal value
R_s	0.436
R_r	0.8160
M	0.0693
$L_s = L_r$	0.0713

Then, the inputs ($V_{s\alpha}$,$V_{s\beta}$)and outputs ($i_{s\alpha}, i_{s\beta}, \Omega_r$)of the MI were measured with a sampling time of $10^{-4}s$and polluted with a white noise signal. The signals obtained are shown inFig7 and 8. These signals are used to construct the regressorΨ described by equation (9) using TRAPEZ integration.

Next,the linear KALMAN filter algorithm, which is discussed in Section 3, is used to estimate the coefficients θ. A description of the estimation technique is given in Fig.

The results obtained are shown inFig30 to 37. The simulation results for the estimated and nominal values of the coefficients θ_1 à $\theta_7$are shown in Fig30 à 37. The latter indicates that the estimates converge to the nominal steady-state values with little estimation error.

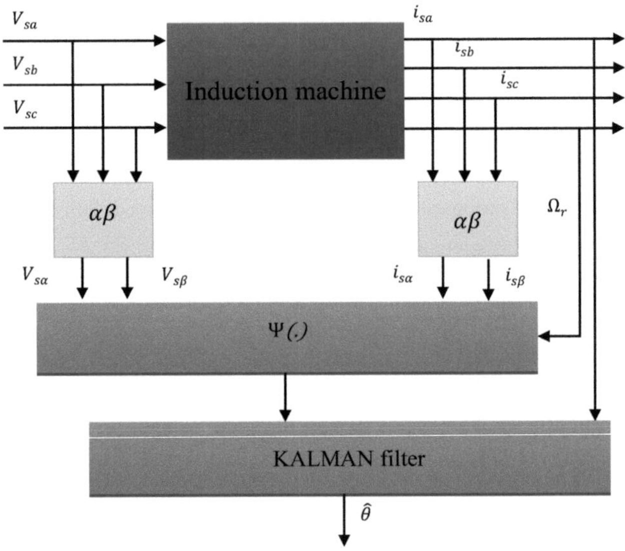

Fig .27.KALMAN filter implementation for parameter identification

By estimating the coefficients θ, we obtain the estimated parameters $\hat{R}_s, \hat{R}_r, \hat{L}_s, \hat{L}_r andM$ using the equations in section 2. The results obtained are shown in the table below.

Table 10. Comparison between nominal and estimated values s

Parameter	Nominal value	Estimated value	Estimation error
R_s	0.436	0.4342	0.4128%
R_r	0.8160	0.8166	0.0735%
$L_s = L_r$	0.0713	0.07433	4.2496%
M	0.0693	0.7229	4.3290%

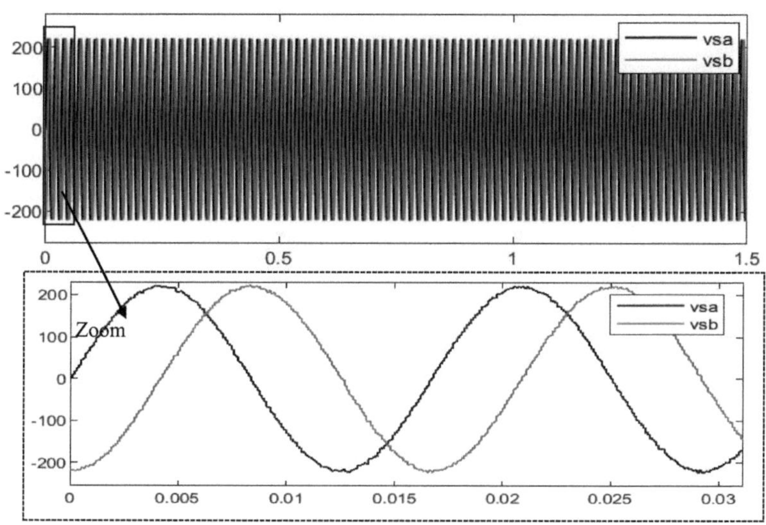

Fig .28. Measurement of stator voltage Vsa Vsb

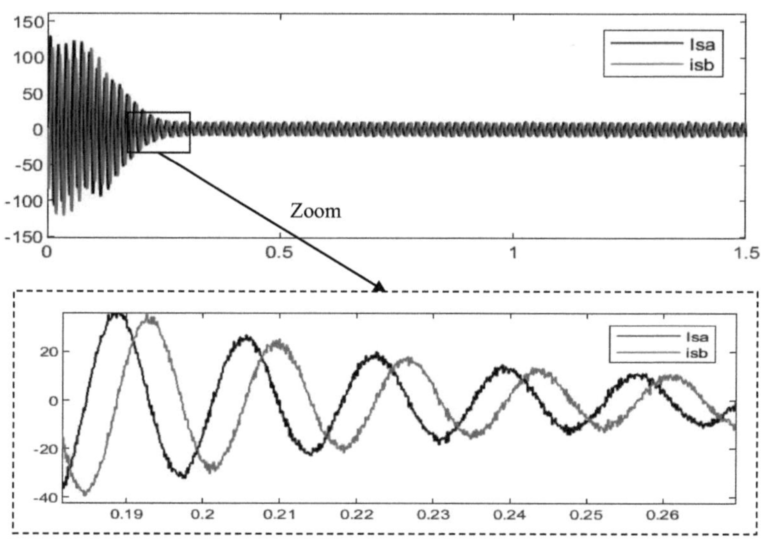

Fig.29.Isa Isb stator current measurements

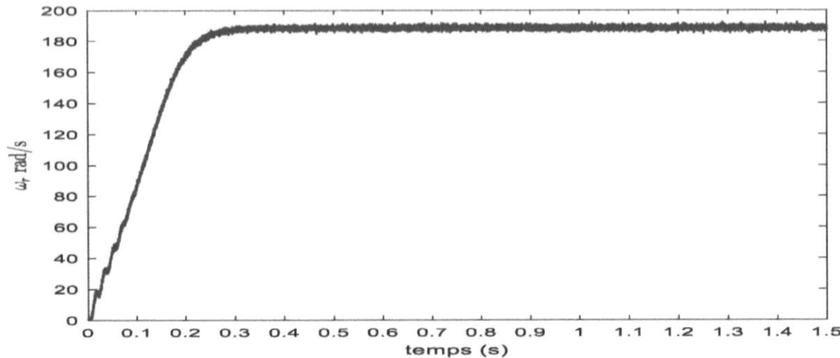

Fig.30.Rotor speed measurement

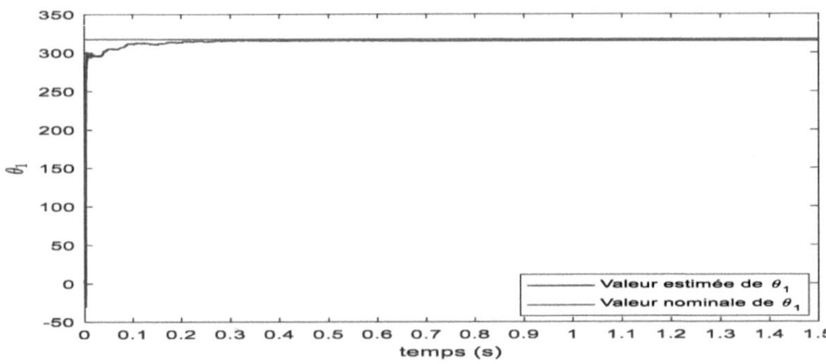

Fig .31.Comparison between the estimated and nominal values of the parameter θ_1

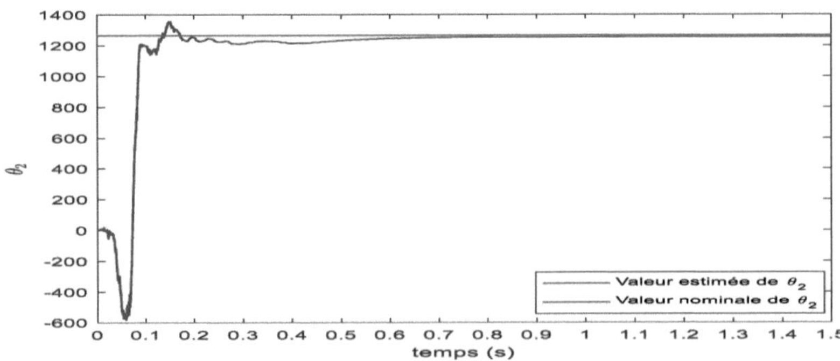

Fig.32.Comparison between the estimated and nominal values of the parameter θ_2

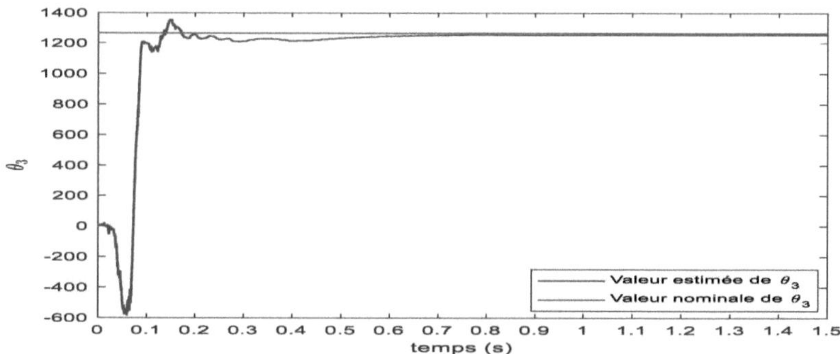

Fig.33.Comparison between the estimated and nominal values of the parameter θ_3

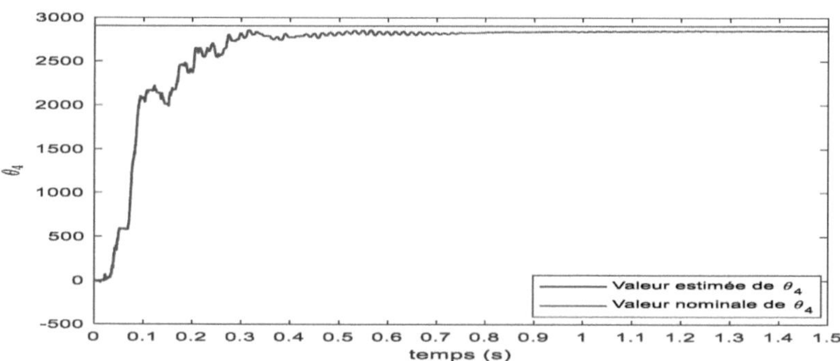

Fig.34.Comparison between the estimated and nominal values of the parameter θ_4

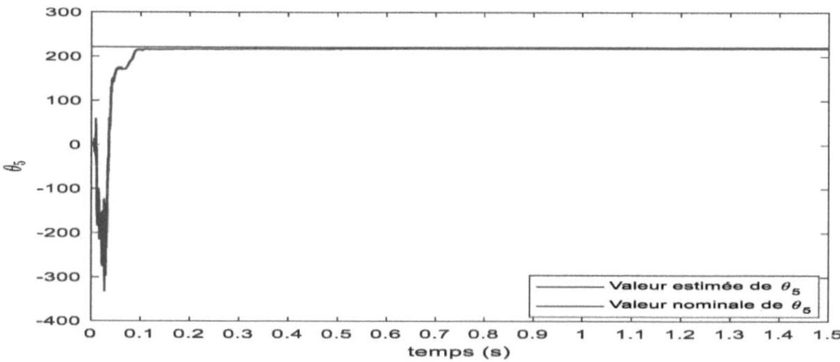

Fig.35.Comparison between the estimated and nominal values of the parameter θ_5

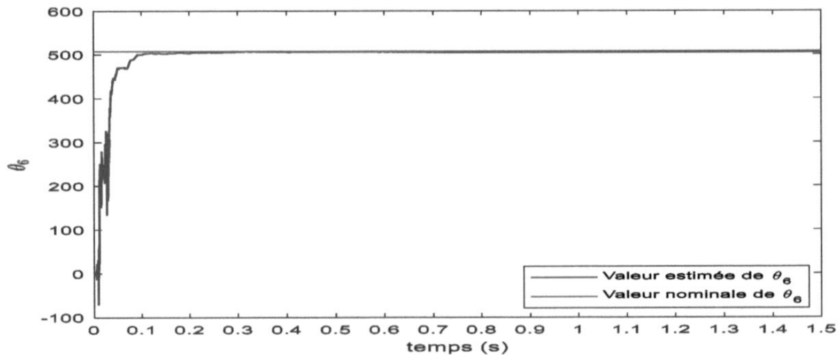

Fig.36.Comparison between the estimated and nominal values of the parameter θ_6

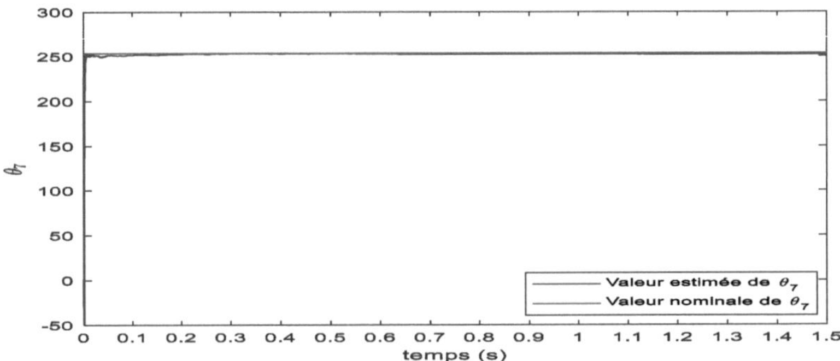

Fig.37. Comparison between the estimated and nominal values of the parameter θ_7

I.2.7. Conclusion of the chapter

In this work, a method based on the KALMAN filter algorithm is used to identify the electrical parameters of an induction motor using the stator currents and voltages, and the rotor speed. To do this, these measured signals were used to develop a regressor, which was used with the KALMAN filter to make the estimate. The results show that the estimated parameters obtained are close to the reference values for the MI considered. This demonstrates the reliability of the proposed identification method for estimating induction motor parameters.

Chapter 2 references

[1] H. Ouadi, A. Barra, and K. El Majdoub, "NONLINEAR CONTROL FOR GRID CONNECTED WIND ENERGY SYSTEM WITH MULTILEVEL INVERTER," vol. 12, no. 4, 2017, Accessed: Oct. 05, 2022. [Online]. Available: www.arpnjournals.com

[2] R. Padilha, R. Zelir, C. Cauduro, and H. Abilio, "Electrical Parameter Identification of Single-Phase Induction Motor by RLS Algorithm," *Induction Mot. - Model. Control*, 2012, doi: 10.5772/37664.

[3] B. Reddy, G. Poddar, and B. P. Muni, "Parameter Estimation and Online Adaptation of Rotor Time Constant for Induction Motor Drive," *IEEE Trans. Ind. Appl.* vol. 58, no. 2, pp. 1416-1428, 2022, doi: 10.1109/TIA.2022.3141700.

[4] F. Karami, J. Poshtan, and M. Poshtan, "Detection of broken rotor bars in induction motors using nonlinear Kalman filters," *ISA Trans.* vol. 49, no. 2, pp. 189-195, 2010, doi: 10.1016/j.isatra.2009.11.005.

[5] S. Ayasun and C. O. Nwankpa, "Induction motor tests using MATLAB/Simulink and their integration into undergraduate electric machinery courses," *IEEE Trans. Educ*, vol. 48, no. 1, pp. 37-46, 2005, doi: 10.1109/TE.2004.832885.

[6] M. Aminu, P. K. Ainah, M. Abana, and U. A. Abu, "Identification of induction machine parameters using only no-load test measurements," *Niger. J. Technol*, vol. 37, no. 3, pp. 742-748, Jul. 2018, doi: 10.4314/njt.v37i3.25.

[7] S. Sehra, K. K. Gautam, and V. Bhuria, "PERFORMANCE EVALUATION OF THREE PHASE INDUCTION MOTOR BASED ON NO LOAD AND BLOCKED ROTOR TEST USING MATLAB".

[8] M. Yazdani-Asrami *et al*, "Parameters Estimation Tests of Induction Machine Using Matlab/Simulink," *J. Phys. Conf. Ser.* 1973, no. 1, p. 012109, Aug. 2021, doi: 10.1088/1742-6596/1973/1/012109.

[9] F. L. Mapelli, A. Bezzolato, and D. Tarsitano, "A rotor resistance MRAS estimator for induction motor traction drive for electrical vehicles," *Proc. - 2012 20th Int. Conf. Electr. Mach. ICEM 2012*, pp. 823-829, 2012, doi: 10.1109/ICELMACH.2012.6349972.

[10] T. M. Rowan, R. J. Kerkman, and D. Leggate, "A Simple On-Line Adaption for Indirect Field Orientation of an Induction Machine," *IEEE Trans. Ind. Appl.* vol. 27, no. 4, pp. 720-727, 1991, doi: 10.1109/28.85488.

[11] H. Madadi Kojabadi, "Active power and MRAS based rotor resistance identification of an IM drive," *Simul. Model. Pract. Theory*, vol. 17, no. 2, pp. 376-389, Feb. 2009, doi: 10.1016/J.SIMPAT.2008.09.014.

[12] M. Munshi and S. G. Choudhuri, "Model Reference Adaptive System using Rotor Flux and Back Emf techniques for speed estimation of an Induction Motor operated in Vector Control mode: A comparative study," *2016 IEEE Uttar Pradesh Sect. Int. Conf. Electr. Comput. Electron. Eng. UPCON 2016*, pp. 44-49, Apr. 2017, doi: 10.1109/UPCON.2016.7894622.

[13] X. Yu, M. W. Dunnigan, and B. W. Williams, "A novel rotor resistance identification method for an indirect rotor flux-orientated controlled induction machine system," *IEEE Trans. Power Electron*, vol. 17, no. 3, pp. 353-364, May 2002, doi: 10.1109/TPEL.2002.1004243.

[14] T. Orlowska-Kowalska, "Application of extended Luenberger observer for flux and rotor time-constant estimation in induction motor drives," *IEE Proc. D Control Theory Appl.*, vol. 136, no. 6, pp. 324-330, 1989, doi: 10.1049/IP-D.1989.0043/CITE/REFWORKS.

[15] T. Du, P. Vas, and F. Stronach, "Design and application of extended observers for joint

state and parameter estimation in high-performance AC drives," *IEE Proc. Electr. Power Appl.* vol. 142, no. 2, pp. 71-78, Mar. 1995, doi: 10.1049/IP-EPA:19951701.

[16] L. C. Zai, C. L. DeMarco, and T. A. Lipo, "An Extended Kalman Filter Approach to Rotor Time Constant Measurement in PWM Induction Motor Drives," *IEEE Trans. Ind. Appl.* vol. 28, no. 1, pp. 96-104, 1992, doi: 10.1109/28.120217.

[17] F. J. Lin, "Robust speed-controlled induction-motor drive using EKF and RLS estimators," *IEE Proc. Electr. Power Appl*, vol. 143, no. 3, pp. 186-192, 1996, doi: 10.1049/IP-EPA:19960287.

[18] J. Campbell and M. Sumner, "Practical sensorless induction motor drive employing an artificial neural network for online parameter adaptation," *IEE Proc. Electr. Power Appl*, vol. 149, no. 4, pp. 255-260, Jul. 2002, doi: 10.1049/IP-EPA:20020289.

[19] D. Fodor, G. Griva, and F. Profumo, "Compensation of parameters variations in induction motor drives using a neural network," *PESC Rec. - IEEE Annu. Power Electron. Spec. Conf.* vol. 2, pp. 1307-1311, 1995, doi: 10.1109/PESC.1995.474983.

[20] B. Karanayil, M. F. Rahman, and C. Grantham, "Online stator and rotor resistance estimation scheme using artificial neural networks for vector controlled speed sensorless induction motor drive," *IEEE Trans. Ind. Electron*, vol. 54, no. 1, pp. 167-176, Feb. 2007, doi: 10.1109/TIE.2006.888778.

[21] S. M. Gadoue, D. Giaouris, and J. W. Finch, "MRAS sensorless vector control of an induction motor using new sliding-mode and fuzzy-logic adaptation mechanisms," *IEEE Trans. Energy Convers.* vol. 25, no. 2, pp. 394-402, Jun. 2010, doi: 10.1109/TEC.2009.2036445.

[22] N. Jabbour and C. Mademlis, "Online Parameters Estimation and Autotuning of a Discrete-Time Model Predictive Speed Controller for Induction Motor Drives," *IEEE Trans. Power Electron*, vol. 34, no. 2, pp. 1548-1559, Feb. 2019, doi: 10.1109/TPEL.2018.2831459.

[23] H. R. Mohammadi and A. Akhavan, "Parameter Estimation of Three-Phase Induction Motor Using Hybrid of Genetic Algorithm and Particle Swarm Optimization," *J. Eng. (United Kingdom)*, vol. 2014, 2014, doi: 10.1155/2014/148204.

[24] D. C. Huynh and M. W. Dunnigan, "Parameter estimation of an induction machine using advanced particle swarm optimisation algorithms," *IET Electr. Power Appl*, vol. 4, no. 9, pp. 748-760, Nov. 2010, doi: 10.1049/IET-EPA.2009.0296/CITE/REFWORKS.

[25] M. A. Awadallah, "Parameter Estimation of Induction Machines from Nameplate Data Using Particle Swarm Optimization and Genetic Algorithm Techniques," http://dx.doi.org/10.1080/15325000801911393, vol. 36, no. 8, pp. 801-814, Aug. 2008, doi: 10.1080/15325000801911393.

[26] S. H. Modélisation, "Modélisation , observation et commande de la machine asynchrone Sofien Hajji To cite this version : HAL Id : tel-01058792 TITLE : Mod '," 2014.

[27] M. Salah-eddine, S. Said, and B. Bahloul, "Parameter Identification of Squirrel Cage Induction Machine using Linear Kalman Filter," pp. 1-6.

[28] Y. S. Mohamed, B. M. Hasaneen, A. A. Elbaset, and A. E. Hussein, "RECURSIVE LEAST SQUARE ALGORITHM FOR ESTIMATING PARAMETERS OF AN INDUCTION MOTOR," *J. Eng. Sci.* 39, no. 1, pp. 87-98, 2011.

[29] J. Tang, Y. Yang, F. Blaabjerg, J. Chen, L. Diao, and Z. Liu, "Parameter identification of inverter-fed induction motors: A review," *Energies*, vol. 11, no. 9, pp. 1-21, 2018, doi: 10.3390/en11092194.

[30] Y. Koubaa, "Recursive identification of induction motor parameters," *Simul. Model. Pract. Theory*, vol. 12, no. 5, pp. 363-381, Aug. 2004, doi:

10.1016/J.SIMPAT.2004.04.003.

[31] P. Dipanjali Behera *et al*, "On-Line Parameter Identification of a Squirrel Cage Induction Motor," *J. Phys. Conf. Ser.* vol. 1302, no. 2, p. 022054, Aug. 2019, doi: 10.1088/1742-6596/1302/2/022054.

[32] H. Zhang, S. J. Gong, and Z. Z. Dong, "On-line parameter identification of induction motor based on RLS algorithm," in *2013 International Conference on Electrical Machines and Systems, ICEMS 2013*, 2013, pp. 2132-2137. doi: 10.1109/ICEMS.2013.6713208.

[33] L. Yang, X. Peng, and Z. Li, "Induction motor electrical parameters identification using RLS estimation," *2010 Int. Conf. Mech. Autom. Control Eng. MACE2010*, pp. 3294-3297, 2010, doi: 10.1109/MACE.2010.5535653.

[34] T. Kostal and P. Kobrle, "Induction machine on-line parameter identification for resource-constrained microcontrollers based on steady-state voltage model," *Electron.* vol. 10, no. 16, 2021, doi: 10.3390/electronics10161981.

[35] F. Debbabi, A. L. Nemmour, A. Khezzar, and S. E. Chelli, "An approved superiority of real-time induction machine parameter estimation operating in self-excited generating mode versus motoring mode using the linear RLS algorithm: Ideas & applications," *Int. J. Electr. Power Energy Syst.* vol. 118, no. November 2019, pp. 105725, 2020, doi: 10.1016/j.ijepes.2019.105725.

[36] M. Boufadene, "Modeling and Control of AC Machine using MATLAB®/SIMULINK," *Model. Control AC Mach. using MATLAB®/SIMULINK*, no. December 2018, pp. 1-50, 2018, doi: 10.1201/9780429029653.

[37] L. Guo, "Estimating time-varying parameters by the kaiman filter based algorithm: Stability and convergence," *IEEE Trans. Automat. Contr.* 35, no. 2, pp. 141-147, 1990, doi: 10.1109/9.45169.

[38] A. Isaksson, "Identification of Time Varying Systems Through Adaptive Kalman Filtering," *IFAC Proc. Vol.* 20, no. 5, pp. 305-310, 1987, doi: 10.1016/s1474-6670(17)55517-2.

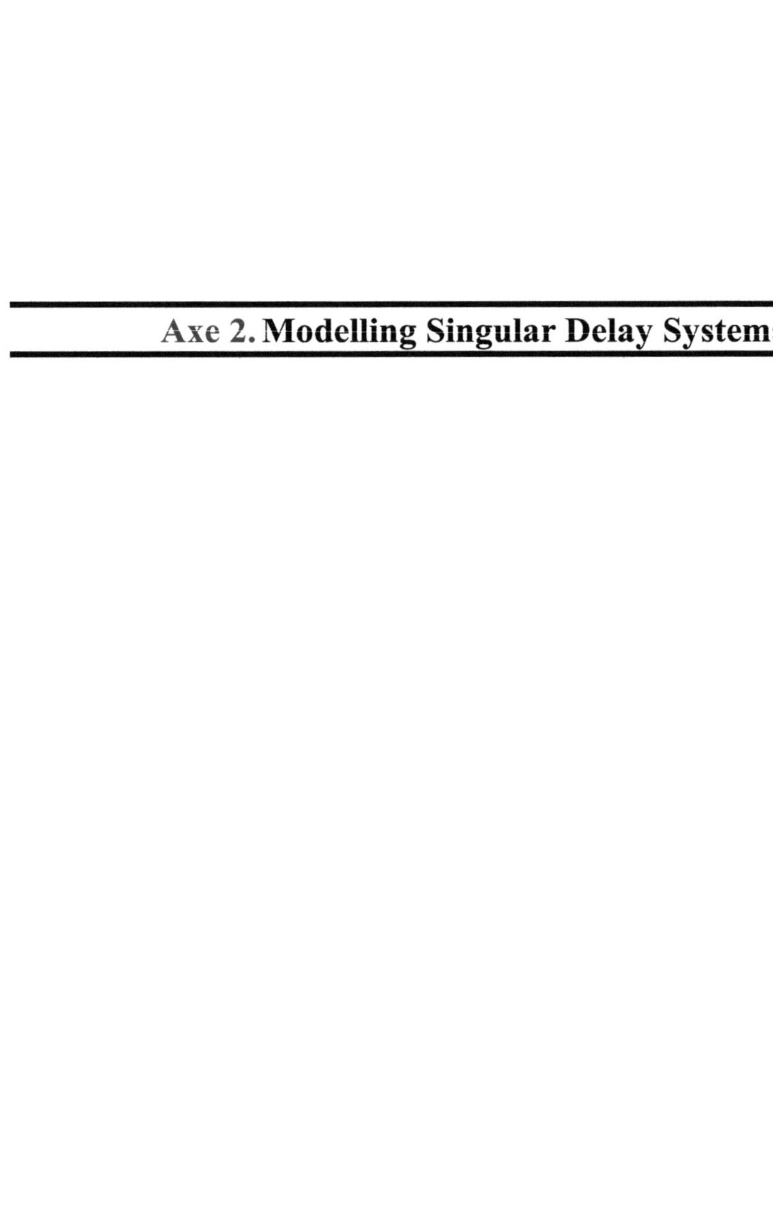

Axe 2. Modelling Singular Delay Systems

Chapter III: Admissibility of singular systems with time-varying delay

---

Summary of chapter III:

This chapter aims to establish and analyse the stability of continuous linear descriptor systems with time-varying delay, based on the well-known Lyapunov KRASOVSKI function method and the linear matrix inequality (LMI) technique. The grounded theorem is used to verify not only stability but also admissibility, so that the regularity and impulse-free nature of the descriptor system under study is assumed.

The established criterion was satisfied using a neutral approach, the transformation of the system, which made it possible to verify the stability of the singular time-varying system via a less complicated calculation. In addition to this new form, a major integral inequality lemma was also used, due to its conservative results. The originality of this contribution has been achieved by combining the double inequality technique and the neutral approach to perform the admissibility analysis of a time-varying delay singular system, the objective of which is to provide an alternative delay-dependent criterion in terms of (LMI), the verification of the proposed theorem has been done through a linear matrix inequality programmed on the MATLAB numerical computation platform.

At the end of this chapter, a numerical example will be provided using the LMI toolbox, in order to prove the validity and efficiency of the proposed method. A comparison table will be presented to show the results developed in other works in relation to the proposed theorem.

II.3.1.Introduction of the chapter :

Analysing the stability of physical systems is one of the main concerns in the field of automation. For this analysis to be successful, so-called physical systems must first be modelled by a mathematical state representation. However, two frequent representations have been widely treated in the literature: the state representation of a regular system [1]and the state representation of a singular system [2].

Studies carried out in the field of automation have shown that the representation of singular systems (also known as systems with a descriptor state space) describes physical models better than the representation of regular systems.

Hence the massive development of work dedicated to this axis by researchers, whose objective is to obtain less conservative results while giving values of delays higher than those existing in previous work, and more effective.

It is well known that a system can be stable with a constant delay, but if the system experiences time-varying delays, this can ruin its stability. This explains the attention given to this subject.

Taking into account the "time-varying delay" factor makes it possible to bypass the stability study of the system's evolution in the future, based on its current state and its previous states. These delays can be seen in real engineering systems, appearing in the state vector, the measurement or even the input.

The influence of this factor can be positive on the system's performance, if it achieves stability, or negative if it reduces its performance, triggering vibrations.

In recent years, many research articles have dealt with the analysis of the stability of physical systems:

- Regular or singular system with constant delay, in continuous or discrete domain[[3]-[4].
- Regular or singular systems with time-varying delay, in continuous or discrete domain[[5]-[[6].

This research has shown that the time-invariant delay criterion is more conservative if the delay "τ" is known, or insignificant. However, these time delays can add up as the signal is transmitted from one point to another.

Consequently, it is preferable to work with a time-varying delay in the state vector, usually denoted by τ (t), or h (t).

In this chapter, we will deal with the admissibility analysis of the singular time-varying delay system, using the widespread Lyapunov-Krasovski method[7][-[[8] and the linear inequality matrix (LMI) technique [9]-[10].

Previous scientific work has shown that the use of various inequalities has facilitated the calculation of Lyapunov derivative functions. These inequalities include Jensen's integral inequality [11]Wirtinger's integral inequality[12]etc.

The originality of this work lies in the implementation of the double integral inequality technique in the admissibility analysis of the singular system with time-varying delay.

The adoption of this inequality was favoured because of its conservative results compared with the other existing inequalities mentioned above.

The first part of the chapter will be devoted to a description of the system under study, with the development of the neutral approach used.

The proposed theorem and the study of stability will be developed in the second part, where the calculation of Lyapunov functional derivatives is illustrated using the double integral inequality technique.

An academic example will be provided in the third part, with a comparison of previous work to show the effectiveness of the suggested theorem.

At the end, a conclusion will be presented, summarising the work done in this chapter.

II.3.2.Problem formulation :

A. Description of the problem:

Descriptor systems with two time-varying additive delays are described by the following representation:

$$\begin{cases} E\dot{x}(t) = Ax(t) + A_d x(t-\tau(t)) \\ x(t) = \delta(t), \qquad t \in [-\tau(t), 0] \end{cases} \qquad (III.\ 1)$$

x(t) is the state vector defined in $\mathbb{R}^n$E (singular matrix),A, B, C and D describe real, constant matrices of suitable dimensions with the state, control and output vectors,τ (t) is a time-varying delay, satisfying the condition

$\forall$ t > 0.

- **Assumption 1:**

In this work we make the following two assumptions:

$$0 \le \tau(t) \le \bar{\tau} \qquad (III.\ 2)$$

$$0 \le \dot{\tau(t)} \le \bar{d} \qquad (III.\ 3)$$

$\bar{\tau}$is a given limit of the delay, $\bar{d}$ is the delay-dependent time derivative .

The lemmas and definitions mentioned in Chapter 2 were used to create the main results.

B. Model transformation :

It is well known that a neutral system is not equivalent to the descriptor system. However, the approach is exploited so that the asymptotic stability of the first system guarantees the admissibility of the singular system and vice versa. To this end, the neutral representation will be used in this work to check the stability of the singular system.

$$\begin{cases} \dot{\mu}(t) - \hat{C}(t)\dot{\mu}(t-h(t)) = \hat{A}\mu(t) + \hat{A}_d\mu(t-h(t)) \\ \mu(t) = \varphi(t) \qquad t\in[-h(t), 0] \end{cases} \qquad (III.\ 4)$$

The following representation is defined for this purpose:

Where the matrices declared are :

$$\hat{A}=\begin{bmatrix}A_1 & 0\\ 0 & -I_{n-r}\end{bmatrix};\quad \hat{A}_d=\begin{bmatrix}A_{d_1} & A_{d_2}\\ -A_{d_3} & -A_{d_4}\end{bmatrix};\quad \hat{C}(t)=\left(1-\dot{h}(t)\right)\begin{bmatrix}0 & 0\\ -A_{d_3} & -A_{d_4}\end{bmatrix};\qquad \text{(III. 5)}$$

i. **Proof of transformation of model:**

If the pair (E,A) is regular and impulse-free, then there are two invertible matrices $M\in\mathbb{R}^{n*n}$ and $N\in\mathbb{R}^{n*n}$ such that :

$$\text{MEN}=\begin{bmatrix}Ir & 0\\ 0 & 0\end{bmatrix}\qquad \text{MAN}=\begin{bmatrix}A_1 & 0\\ 0 & I_{n-r}\end{bmatrix}\qquad \text{(III. 6)}$$

According to this lemma and with the aim of creating the neutral approach, we can rewrite the previous terms as follows.

$$\text{MEN}=\bar{E}\qquad \bar{E}=\begin{bmatrix}Ir & 0\\ 0 & 0\end{bmatrix}\qquad \text{(III.7)}$$

$$\text{MAN}=\bar{A}\qquad \bar{A}=\begin{bmatrix}A_1 & 0\\ 0 & I_{n-r}\end{bmatrix}\qquad \text{(III. 8)}$$

It is assumed that:

$$MA_dN=\bar{A}_d=\begin{bmatrix}A_{d_1} & A_{d_2}\\ A_{d_3} & A_{d_4}\end{bmatrix}\qquad N^{-1}x(t)=\mu(t)=\begin{bmatrix}\mu_1(t)\\ \mu_2(t)\end{bmatrix}\qquad \text{(III. 9)}$$

If the dimensions are compatible with $\bar{E}$ the singular system becomes equivalent to :

$$\bar{E}\dot{\mu}(t)=\bar{A}\mu(t)+\bar{A}_d\mu(t-h(t))\qquad \text{(III. 10)}$$

Using assumption (III.10), expression (III. 11) can be rewritten as :

$$\begin{cases}\dot{\mu}_1(t)= A_1\mu_1(t)+A_{d_1}\mu_1(t-h(t))+A_{d_2}\mu_2(t-h(t))\\ 0=\mu_2(t)+A_{d_3}\mu_1(t-h(t))+A_{d_4}\mu_2(t-h(t))\end{cases}\qquad \text{(III. 12)}$$

After deriving the second expression of (III.7) we get:

$$\frac{d}{dt}[\mu_2(t)+A_{d_3}\mu_1(t-h(t))+A_{d_4}\mu_2(t-h(t))]=0\qquad \text{(III. 13)}$$

Merging expressions (III.7) and (III.8) we find **:**

$$\begin{aligned}\dot{\mu}_2(t)=-\left(1-\dot{h}(t)\right)A_{d_3}\dot{\mu}_1(t-h(t))-\left(1-\dot{h}(t)\right)A_{d_4}\dot{\mu}_2(t-h(t))\\ -\mu_2(t)-A_{d_3}\mu_1(t-h(t))-A_{d_4}\mu_2(t-h(t))\end{aligned}\qquad \text{(III.14)}$$

The new system can therefore be deduced as follows:

$$\begin{bmatrix}\dot{\mu}_1(t)\\ \dot{\mu}_2(t)\end{bmatrix}=\begin{bmatrix}A_1\mu_1(t)+A_{d_1}\mu_1(t-h(t))+A_{d_2}\mu_2(t-h(t))\\ -\mu_2(t)-A_{d_3}\mu_1(t-h(t))+A_{d_4}\mu_2(t-h(t))\end{bmatrix}+(1-\qquad \text{(III.15)}$$

$$\dot{h}(t))\begin{bmatrix}0 & 0\\ -A_{d_3} & -A_{d_4}\end{bmatrix}\begin{bmatrix}\dot{\mu}_1(t-h(t))\\ \dot{\mu}_2(t-h(t))\end{bmatrix}$$

We define the matrices $\widehat{A}$, $\widehat{A}_d$and $\widehat{C}(t)$ by (III.5):

$$\widehat{A}=\begin{bmatrix}A_1 & 0\\ 0 & -I_{n-r}\end{bmatrix};\qquad \widehat{A}_d=\begin{bmatrix}A_{d_1} & A_{d_2}\\ -A_{d_3} & -A_{d_4}\end{bmatrix};\qquad \widehat{C}(t)=(1-\dot{h}(t))\begin{bmatrix}0 & 0\\ -A_{d_3} & -A_{d_4}\end{bmatrix}$$

The representation of the neutral system is therefore (III.4) . Based on this approach, we seek to establish the new criterion for verifying the stability of the singular system dependent on time-varying delays.

II.3.3.Stability study :

The study of the stability of the singular system has been addressed in several previous research projects, using existing methods in the literature.

Since the emergence of the neutral approach [13]stability analysis has become easier. In this section, a new theorem will be proposed, together with a precise development explaining the results obtained.

A. Theorem :

For a $\bar{\tau}$ the neutral system is asymptotically stable, proving the stability of the singular system. If there is a positive matrix P of dimension 3n*3n, where $P=P^T$n*n positive matrix Q_i (i=1,2) positive, with $Q_i = {Q_i}^T$and $R^T=R > 0$.

The next linear matrix inequality is: ζ< 0

Where:

$$\zeta=\begin{bmatrix}\zeta_{11} & \zeta_{12} & \zeta_{13} & \zeta_{14} & \zeta_{15}\\ * & \zeta_{22} & \zeta_{23} & \zeta_{24} & \zeta_{25}\\ * & * & \zeta_{33} & \zeta_{34} & \zeta_{35}\\ * & * & * & \zeta_{44} & \zeta_{45}\\ * & * & * & * & \zeta_{55}\end{bmatrix} \qquad \text{(III. 16)}$$

The elements of the matrixζ are :

$\zeta_{11}=P_{11}A_1+2P_{13}-4R+Q+h^2{A^T}_1RA_1+{A^T}_1SA_1$	$\zeta_{14}=P_{33}-\frac{24R}{\bar{\tau}}$
$\zeta_{12}=P_{11}C_1+P_{12}+3R+{A^T}_1S\widehat{A}_d+h^2{A^T}_1R\widehat{A}_d$	$\zeta_{15}=\frac{60R}{\bar{\tau}}$
$\zeta_{13}=-P_{12}+P_{23}+P_{11}\widehat{A}_d+A_1P_{12}+{A^T}_1SC_1+h^2{A^T}_1RC_1$	$\zeta_{24}=-P_{33}+\frac{36R}{\bar{\tau}}$
$\zeta_{22}=P_{12}\widehat{A}_d-2P_{23}-Q+{\widehat{A}_d}^TS\widehat{A}_d-9R+h^2{\widehat{A}_d}^TR\widehat{A}_d$	$\zeta_{25}=\frac{-60R}{\bar{\tau}^2}$

ζ_{23}=$P_{11}C_1+P_{22}$+$\hat{A}_d{}^T$SC_1+$\hat{A}_d{}^T$RC_1	ζ_{35}= 0
ζ_{33}= $C^T{}_1SC_1$-S+$h^2C^T{}_1$RC_1	ζ_{44}=$-\frac{192R}{\bar{\tau}^2}$
ζ_{34}=$P_{23}+C^T{}_1P_{13}$	ζ_{45}=$\frac{360R}{\bar{\tau}^3}$
ζ_{55}=$-\frac{720R}{\bar{\tau}^4}$	

B. Proof of the theorem:

For the sake of computational complexity, let's define the following matrices:

σ(t)=$[\mu^T(t)\mu^T(t-\tau(t))\int_{t-\tau(t)}^{t}\mu^T(s)ds]^T$

$$\chi^T(t) = \left[\mu^T(t) \quad \mu^T(t-\tau(t)) \quad \dot{\mu}^T(t-\tau(t)) \quad \int_{t-\tau(t)}^{t}\mu(s)^T ds \quad \int_{-\tau(t)}^{t}\int_{u}^{t}\mu(s)^T dsdu\right] \quad \text{(III. 17)} \quad \text{(III. 18)}$$

To prove the stability of the singular system (III.1) , it is sufficient to verify the stability of the neutral system (III.4), since stability is equivalent between the two systems .

The Lyapunov Krasovski function chosen for this new criterion is :

$$V(\mu_t) = V_1(\mu_t) + V_2(\mu_t) + V_3(\mu_t) + V_4(\mu_t) \quad \text{(III. 19)}$$

The four functionals that form the Lyapunov candidate function are defined as follows:

$$V_1(\mu_t) = \sigma^T(t)P\,\sigma(t) \quad \text{(III. 20)}$$

$$V_2(\mu_t) = \int_{t-\tau(t)}^{t}\mu^T(s)Q\mu(s)ds \quad \text{(III. 21)}$$

$$V_3(\mu_t) = \int_{t-\tau(t)}^{t}\dot{\mu}(s)^T S\dot{\mu}(s)\,ds \quad \text{(III. 22)}$$

$$V_4(\mu_t) = \tau\int_{-\tau(t)}^{0}\int_{t+\theta}^{t}\dot{\mu}^T(s)R\dot{\mu}(s)dsd\theta \quad \text{(III. 23)}$$

Using the Newton-Leibniz model transformation, the derivation of $V_1(\mu_t)$gives :

$$\dot{V}_1(\mu_t) = 2\,[\mu^T(t)\mu^T(t-\tau(t))\int_{t-\bar{\tau}}^{t}\mu^T(s)ds]\begin{bmatrix}P_{11} & P_{12} & P_{13}\\ * & P_{22} & P_{23}\\ * & * & P_{33}\end{bmatrix}\begin{bmatrix}\dot{\mu}(t)\\ \dot{\mu}(t-\tau(t))\\ \int_{t-\bar{\tau}}^{t}\dot{\mu}(s)ds\end{bmatrix} \quad \text{(III.24)}$$

$$=2\sigma^T(t)\begin{bmatrix} P_{11} & P_{12} & P_{13} \\ * & P_{22} & P_{23} \\ * & * & P_{33} \end{bmatrix}\begin{bmatrix} \hat{A}\mu(t)+\hat{A}_d\mu(t-\tau(t))+\hat{C}\dot{\mu}(t-\tau(t)) \\ \dot{\mu}(t-\tau(t)) \\ \mu(t)-\mu(t-\tau(t)) \end{bmatrix} \quad \text{(III.25)}$$

$$\dot{V}_1(\mu_t)=\begin{bmatrix} \mu(t) \\ \mu(t-\tau(t)) \\ \dot{\mu}(t-\tau(t)) \\ \int_{t-\tau(t)}^{t}\mu(s)ds \\ \int_{-\tau(t)}^{t}\int_{u}^{t}\mu(s)dsdu \end{bmatrix}^T \begin{bmatrix} \alpha_{11} & \alpha_{12} & \alpha_{13} & \alpha_{14} & \alpha_{15} \\ * & \alpha_{22} & \alpha_{23} & \alpha_{34} & \alpha_{25} \\ * & * & \alpha_{33} & \alpha_{34} & \alpha_{35} \\ * & * & * & \alpha_{44} & \alpha_{45} \\ * & * & * & * & \alpha_{55} \end{bmatrix}\begin{bmatrix} \mu(t) \\ \mu(t-\tau(t)) \\ \dot{\mu}(t-\tau(t)) \\ \int_{t-\tau(t)}^{t}\mu(s)ds \\ \int_{-\tau(t)}^{t}\int_{u}^{t}\mu(S)dsdu \end{bmatrix} \quad \text{(III.26)}$$

The elements of theα matrix are:

$\boldsymbol{\alpha_{11}=P_{11}\hat{A}+\hat{A}^TP_{11}+2P_{13}}$	$\alpha_{12}=P_{11}\hat{A}_d+\hat{A}P_{12}+P_{23}-P_{13}$
$\boldsymbol{\alpha_{13}=P_{11}\hat{C}+P_{12}}$	$\alpha_{14}=P_{33}$
$\boldsymbol{\alpha_{15}}=0$	$\alpha_{22}=P_{12}\hat{A}_d+\hat{A}^T{}_dP_{12}-2P_{23}$
$\boldsymbol{\alpha_{23}=P_{12}\hat{C}+P_{22}}$	$\alpha_{24}=-P_{33}$
$\boldsymbol{\alpha_{25}}=0$	$\alpha_{33}=0$
$\boldsymbol{\alpha_{34}=\hat{C}^TP_{13}+P_{23}}$	$\alpha_{35}=0$
$\boldsymbol{\alpha_{44}}=0$	$\alpha_{45}=0$
$\boldsymbol{\alpha_{55}}=0$	

The derivation of $V_2(\mu_t)$ gives the result :

$$\dot{V}_2(\mu_t)=\mu^T(t)Q\mu(t)-(1-d)\,\mu^T(t-\tau(t))Q_1\mu(t-\tau(t)) \quad \text{(III. 27)}$$

$$\dot{V}_2(\mu_t)=\chi(t)^T\begin{bmatrix} Q & 0 & 0 & 0 & 0 \\ * & 0 & 0 & 0 & 0 \\ * & * & -(1-d)Q & 0 & 0 \\ * & * & * & 0 & 0 \\ * & * & * & * & 0 \end{bmatrix}\chi(t) \quad \text{(III. 28)}$$

Calculating the derivative of $V_3(\mu_t)$gives the following :

$$\dot{V}_3(\mu_t) = \left(\mu^T(t)\hat{A}^T + \mu^T(t-\tau(t))\hat{A}_d^T + \dot{\mu}^T(t-\tau(t))\hat{C}^T\right) S \quad \text{(III.29)}$$

$$\left(\hat{A}\mu(t) + \hat{A}_d\mu(t-\tau(t)) + \hat{C}\dot{\mu}(t-\tau(t))\right) - (1-d)$$

$$\dot{\mu}^T(t-\tau(t))S\,\dot{\mu}(t-\tau(t))$$

The matrix expression for the derivative of the third function becomes :

$$\dot{V}_3(\mu_t) = \chi(t)^T \begin{bmatrix} \delta_{11} & \delta_{12} & \delta_{13} & \delta_{14} & \delta_{15} \\ * & \delta_{22} & \delta_{23} & \delta_{24} & \delta_{25} \\ * & * & \delta_{35} & \delta_{34} & \delta_{35} \\ * & * & * & \delta_{44} & \delta_{45} \\ * & * & * & * & \delta_{55} \end{bmatrix} \chi(t) \quad \text{(III. 30)}$$

We define the elements of the matrix δ :

δ_{11}=$\hat{A}^T S\hat{A}$	δ_{22}=$\hat{A}_d^T S\hat{A}_d$	δ_{33}=$\hat{C}^T S\hat{C} - (1-d)S$	δ_{44}=0	δ_{55}=0
δ_{12}=$\hat{A}^T S\hat{A}_d$	δ_{23}=$\hat{A}_d^T S\hat{C}$	δ_{34}=$\hat{A}^T Q_2\hat{C}$	δ_{45}=0	
δ_{13}=0	δ_{24}=0	δ_{35}=0		
δ_{14}=0	δ_{25}=0			
δ_{15}=0				

After deriving $V_4(\mu_t)$along the trajectory of the neutral system (III.4) :

$$V_4(\dot{\mu_t}) = h(t)[\dot{\mu}^T(t)R\dot{\mu}(t)]\text{-}\int_{t-h(t)}^{t} \dot{\mu}^T(s)R\dot{\mu}(s)ds \quad \text{(III. 31)}$$

The expression for $\dot{\mu}(t)$ is shown in the following derivative :

$$\dot{V}_4(\mu_t) = \tau^2(t)\left[\hat{A}\,\mu(t) + \hat{A}_d\mu(t-\tau(t)) + \hat{C}\dot{\mu}(t-\tau(t))\right]^T \quad \mathrm{R}\left[\hat{A}\,\mu(t) + \hat{A}_d\mu(t-\tau(t)) + \hat{C}\dot{\mu}(t-\tau(t))\right]\text{-}\tau\int_{t-h(t)}^{t} \dot{\mu}^T(s)R\dot{\mu}(s)ds \quad \text{(III.32)}$$

The result obtained by deriving the fourth function $V_4(\mu_t)$, will be subdivided into two terms, which will be calculated as follows:

- **The first Term** :

$$\Gamma = \tau^2(t)\left[\hat{A}\mu(t) + \hat{A}_d\mu(t-\tau(t)) + \hat{C}\dot{\mu}(t-\tau(t))\right]^T \quad \text{(III. 33)}$$

$$.R[\hat{A}\mu(t)\hat{A}_d\mu(t-\tau(t))+\hat{C}\dot{\mu}(t-\tau(t))]$$

Knowing that τ (t) $\leq \bar{\tau}$ the first term gives the inequality (III:34)

$$\Gamma \leq \overline{\tau^2}\chi(t)^T[\hat{A} \quad \hat{A}_d \quad \hat{C} \quad 0 \quad 0]^T R[\hat{A} \quad \hat{A}_d \quad \hat{C} \quad 0 \quad 0]\chi(t) \qquad \text{(III. 34)}$$

- **The second term:**

$$\Lambda = -\tau\int_{t-h(t)}^{t} \dot{\mu}^T(s)R\dot{\mu}(s)ds \qquad \text{(III. 35)}$$

By applying the double integral inequality technique, and taking into account the inequality (III.2)

The expression Λ will be expanded by

$$\Lambda \leq -\{\tfrac{1}{\bar{\tau}}\left[((\mu(t)-\mu(t-\tau(t)))^T R\left(\mu(t)-\mu(t-\tau(t))\right)\right]+\tfrac{3}{\bar{\tau}}\left[\mu(t)+\mu(t-\tau(t))\tfrac{2}{\bar{\tau}}\int_{t-\tau(t)}^{t}\mu(s)ds\right]^T R\left[\mu(t)+\mu(t-\tau(t)) -\tfrac{2}{\bar{\tau}}\int_{t-\tau(t)}^{t}\mu(s)ds\right]+ \tfrac{5}{\bar{\tau}}\left[\mu(t)+\mu(t-\tau(t))+\tfrac{6}{\bar{\tau}}\int_{t-\tau(t)}^{t}\mu(s)ds - \tfrac{12}{\bar{\tau}^2}\int_{-\tau(t)}^{t}\int_{u}^{t}\mu(S)dsdu\right]^T R\left[\mu(t)+\mu(t-\tau(t))+\tfrac{6}{\bar{\tau}}\int_{t-\tau(t)}^{t}\mu(s)ds-\tfrac{12}{\bar{\tau}^2}\int_{-\tau(t)}^{t}\int_{u}^{t}\mu(S)dsdu\right] \qquad \text{(III.36)}$$

The term found (III.36) can be rewritten in the form:

$$\Lambda \leq \chi(t)^T\begin{bmatrix}\theta_{11} & \theta_{12} & \theta_{13} & \theta_{14} & \theta_{15}\\ * & \theta_{22} & \theta_{23} & \theta_{24} & \theta_{25}\\ * & * & \theta_{35} & \theta_{34} & \theta_{35}\\ * & * & * & \theta_{44} & \theta_{45}\\ * & * & * & * & \theta_{55}\end{bmatrix}\chi(t) \qquad \text{(III. 37)}$$

With :

$\theta_{11} = -4R$	θ_{22}=- 9R	θ_{33}= 0	$\theta_{44}=\frac{-192R}{\bar{\tau}^2}$
θ_{12}=3R	θ_{23}= 0	θ_{34}= 0	$\theta_{45}=\frac{360R}{\bar{\tau}^3}$
θ_{13} =0	$\theta_{24}=\frac{36R}{\bar{\tau}}$	θ_{35}= 0	$\theta_{55}=\frac{-720R}{\bar{\tau}^4}$

$\theta_{14} = \frac{-24}{\bar{\tau}}$	$\theta_{25} = \frac{-60R}{\bar{\tau}^2}$		
$\theta_{15} = \frac{60R}{\bar{\tau}}$			

Combining the two terms (III .34) and (III .37) gives the following inequation:

$$\dot{V}_4(\mu_t) \leq \chi(t)^T \begin{bmatrix} \gamma_{11} & \gamma_{12} & \gamma_{13} & \gamma_{14} & \gamma_{15} \\ * & \gamma_{22} & \gamma_{23} & \gamma_{24} & \gamma_{25} \\ * & * & \gamma_{35} & \gamma_{34} & \gamma_{35} \\ * & * & * & \gamma_{44} & \gamma_{45} \\ * & * & * & * & \gamma_{55} \end{bmatrix} \chi(t) \qquad \text{(III. 38)}$$

We list the elements of γ :

$\gamma_{11} = -4R + \overline{\tau^2}\, \hat{A}^T R \hat{A}$	$\gamma_{22} = -9R + \overline{\tau^2} \hat{A}_d^T R \hat{A}_d$
$\gamma_{12} = 3R + \overline{\tau^2} \hat{A}^T R \hat{A}_d$	$\gamma_{23} = \overline{\tau^2} \hat{A}_d^T R \hat{C}$
$\gamma_{13} = \overline{\tau^2} \hat{A}^T R \hat{C}$	$\gamma_{24} = \frac{36R}{\bar{\tau}}$
$\gamma_{14} = \frac{-24}{\bar{\tau}}$	$\gamma_{45} = \frac{360R}{\bar{\tau}^3}$
$\gamma_{15} = \frac{60R}{\bar{\tau}}$	$\gamma_{33} = \overline{\tau^2} \hat{C}^T R \hat{C}$
$\gamma_{34} = 0$; $\gamma_{35} = 0$	$\gamma_{44} = \frac{-192R}{\bar{\tau}^2}$
$\gamma_{45} = \frac{360R}{\bar{\tau}^3}$	$\gamma_{55} = \frac{-720R}{\bar{\tau}^4}$

The addition of $\dot{V}_1(\mu_t), \dot{V}_2(\mu_t), \dot{V}_3(\mu_t)$ et $\dot{V}_4(\mu_t)$,gives the following linear matrix inequality:

$$\dot{V}(x_t) \leq \begin{bmatrix} \mu(t) \\ \mu(t-\tau(t)) \\ \dot{\mu}(t-\tau(t)) \\ \int_{t-\tau(t)}^{t} \mu(s)ds \\ \int_{-\tau(t)}^{t} \int_{u}^{t} \mu(s)dsdu \end{bmatrix}^T \zeta \begin{bmatrix} \mu(t) \\ \mu(t-\tau(t)) \\ \dot{\mu}(t-\tau(t)) \\ \int_{t-\tau(t)}^{t} \mu(s)ds \\ \int_{-\tau(t)}^{t} \int_{u}^{t} \mu(s)dsdu \end{bmatrix} \quad \text{(III. 39)}$$

The proof of stability is achieved by studying the negativity of the Lyapunov derivative, which can be granted by imposing negative ζ, deducing then that the neutral system is asymptotically stable, so the asymptotic stability of the delayed singular system is proved.

A. Remarks :

- The theorem presented imposes a stability condition on the singular system, with a time-varying delay, using a linear matrix inequality, through the neutral system approach [14].
- The proposed theorem has been developed in this work using a transformation of the neutral approach to simplify the matrix calculations.
- The stability of the proposed singular system is not the only condition for verifying its admissibility.
- The use of the double integral inequality technique in the derivative of the Lyapunov function, generates negative terms, which will allow us to have less conservative results, this comes to explain that this inequality is more accurate than others presented in previous works [15].
- The criteria established enable stability to be verified in the two types of time-varying delay systems, neutral and singular, in the knowledge that they are not equivalent.
- The delay taken into account in this work must be non-zero, because of its existence in the denominator of the linear matrix.
- According to the literature, the application of the double integral inequality technique, which is used to verify the stability of time-varying singular delay systems via a neutral transformation, has not yet been addressed.
- Denoting that the eigenvalues of the matrix $\hat{C}(t)$ are inside the unit circle [42]for this purpose $|\dot{\tau}(t)|$ will be chosen as well <1.

- The results of this new criterion will be compared with some existing criteria, as in [17]-[18]-[12] , so that the theorem found will allow us to obtain better performances.

II.3.4.Numerical example:

B. Example Academic:

In this section, a numerical example will be presented to demonstrate the effectiveness of the proposed criterion. Considering the parameters of the time-varying delay system :

$$A=\begin{bmatrix}-0.5 & 0\\ 0 & -1\end{bmatrix} \qquad E=\begin{bmatrix}1 & 0\\ 0 & 0\end{bmatrix} \qquad A_d=\begin{bmatrix}-1 & 1\\ 0 & 0.5\end{bmatrix} \qquad \text{(III. 40)}$$

Note that, since the delay varies with time, satisfying (III.2) and (III.3), there are two matrices invertible, so the pair (E,A) is regular and impulse-free..:

$$M=\begin{bmatrix}1 & 0\\ 0 & 1\end{bmatrix} \qquad N=\begin{bmatrix}1 & 0\\ 0 & -1\end{bmatrix} \qquad \text{(III. 41)}$$

We pose:

$$MEN=\begin{bmatrix}1 & 0\\ 0 & 0\end{bmatrix} \qquad MAN=\begin{bmatrix}-0.5 & 0\\ 0 & 1\end{bmatrix} \qquad MA_dN=\begin{bmatrix}-1 & 1\\ 0 & -0.5\end{bmatrix} \qquad \text{(III. 42)}$$

Following the neutral approach, we define :

$$\hat{C}(t)=(1-\dot{h}(t))\begin{bmatrix}0 & 0\\ 0 & 0.5\end{bmatrix} \qquad \hat{A}=\begin{bmatrix}-0.5 & 0\\ 0 & -1\end{bmatrix} \qquad \widehat{A_d}=\begin{bmatrix}-1 & 1\\ 0 & 0.5\end{bmatrix} \qquad \text{(III. 43)}$$

Using MATAB's LMI toolbox, with a given derivative d of the singular time-varying delay system, the delay can be deduced so that the system is stable and therefore admissible by the new criteria. Table 1 lists the estimated upper bounds of different works that ensure the stability of descriptor systems. This shows that the criteria developed provide better or identical results to those in [5].

Table 12. Comparative table of upper limits of h for different works

Methods	**The upper bound *h***
Theorem 3 [17]	1
Corollary 1 [18]	1.1547
Theorem 1 [12]	1.1547
Corollary 3.2 [8]	1.1547
Corollary 3.9 [17]	1.1547
Corollary 3.2 [19] (N = 1)	1.1547
Corollary 3.2 [19] (N = 2)	1.1954

Corollary 3.2 [19] (N = 10)	1.2060
Theorem 1 [22]	1.2086
The proposed theorem	1.26043

C. Discussion of the results :

In this chapter, an asymptotic stability of a delayed singular system has been achieved through LMI conditions, combining the neutral approach and the double major integral inequality technique.

The choice of modelling a physical system as a singular system with a time-varying delay was not made arbitrarily, but was considered because of its better stability results.

The stability system was analysed using a Lyapunov Krasovski function chosen on the basis of positive decision matrices.

The theorem delivered was tested on an academic example, and showed good results compared with those in the literature. This is due to the fact that the higher the delay 'h', the more effective the theorem.

II.3.5.Conclusion of the chapter:

This work deals with a new admissibility criterion for the time-varying delay system descriptor [12], based on the combination of two methods which are the transformation of a singular system into a neutral system [13]-[20] and the introduction of the double integral inequality in the calculation of the derivative of the Lyapunov function [19].

The new criterion was proposed in terms of LMI, emphasising that this method is valid for testing the stability of singular systems with variable time delay, as well as for the neutral system with variable time delay, and that this test can be applied to any system (physical, chemical, biological.....)[25].

A numerical example was given at the end of the chapter to show the conservatism of the proposed method. The simulations clearly show that the proposed method has the same or better performance than those compared with, proving the effectiveness of the criteria achieved.

Chapter 3 references

[1]. A. Hmamed, H. El Aiss, and A. EL Hajjaji, "Stability analysis of linear systems with time varying delay: An input output approach," in 2015 54th IEEE Conference on Decision and Control (CDC), Osaka: IEEE, Dec. 2015, p. -17561761. doi: 10.1109/CDC.2015.7402464.

[2]. "2007_Raouf.pdf

[3]. "Lee et al - 2004 - Delay-dependent robust H∞ control for uncertain sy.pdf".

[4]. "Wu and Zhou - 2007 - Delay-dependent Robust Stabilization for Uncertain.pdf".

[5]. "Zhong and Yang - 2008 - DELAY-DEPENDENT ROBUST CONTROL OF DESCRIPTOR SYSTEM.pdf".

[6]. "Liu - 2014 - Improved delay-dependent stability criteria for ne.pdf".

[7]. "Xu et al. - 2017 - Stability analysis of linear systems with two addi.pdf".

[8]. "Fridman - 2002 - Stability of linear descriptor systems with delay.pdf".

[9]. "Boyd - 1994 - Linear matrix inequalities in system and control t.pdf".

[10]. "10.1109@cdc.1994.411440.pdf".

[11]. "Han - 2005 - Absolute stability of time-delay systems with sect.pdf".

[12]. "Seuret and Gouaisbaut - 2013 - Wirtinger-based integral inequality Application t.pdf".

[13]. "Han - 2008 - A Delay Decomposition Approach to Stability of Lin.pdf".

[14]. "Liu - 2014 - Improved delay-dependent stability criteria for ne.pdf".

[15]. "El Haouti et al - 2020 - The employment of neutral approach for linear sing.pdf.

[16]. "Liu et al - 2016 - Admissibility analysis for linear singular systems.pdf".

[17]. "Hmamed et al - 2015 - Stability analysis of linear systems with time var.pdf".

[18]. X. Sun, Q.-L. Zhang, C.-Y. Yang, Z. Su, and Y.-Y. Shao, "An improved approach to delay-dependent robust stabilization for uncertain singular time-delay systems," Int. J. Autom. Comput. vol. 7, no. 2, pp. 205212-, May 2010, doi: 10.1007/s11633-010-0205-5.

Axe 3. Modelling and controlling the order picking process in a warehouse

Chapter IV: Modelling and controlling the order picking process in a warehouse

Summary of chapter IV :

Order picking is the process of retrieving goods from picking locations to fulfil a specific customer order, and is known to be one of the most labour-intensive and costly of all warehouse functions. However, the efficiency of the picking process depends primarily on the manager's decisions, which are often based on intuition and experience. Due to the increasing complexity of the processes, resulting mainly from the volatility of demand and the multitude and uncertainty of operating parameters, these decisions are still far from providing an optimal mode of operation. In this section, a dynamic modelling and control approach for the manual order picking process is proposed in order to perform a sensitivity analysis. First, the model of the order picking process with its performance indicators based on dynamic modelling will be presented by detailing the different components of the system, and then the sensitivity analysis process based on Monte Carlo simulation that will be run on Simulink software will be described in order to locate the parameters that most influence the model outputs. The proposed approach aims to determine the responsiveness of operational processes in meeting defined operational performances, which, for example, helps warehouse managers to make optimal tactical decisions in terms of the means and resources to be implemented in an order-picking activity.

III.4.1. Introduction to the chapter

Warehousing is one of the most important and critical logistics activities in industrial and service systems, in addition to its role in balancing fluctuations between supply and demand and consolidating products for smooth logistics. Warehousing systems have a significant impact on product quality, customer service levels and overall logistics costs. It is therefore important to pay particular attention to warehouses, as they have a significant impact on the efficiency and effectiveness of the entire supply chain. Order picking is one of the most complicated warehousing processes. It is the most expensive activity in a warehouse, accounting for between half and three quarters of the total cost of warehouse operations [1]. In practice, there are two types of order picking system used in a warehouse: the first is the "man-machine" system, in which the order picker moves to the storage location and picks the goods ordered from the picking locations. This type of order-picking process is suitable for slow-moving products, which do not therefore justify major investment. Moving the order picker to the storage locations is an elementary operation that does not necessarily require very sophisticated equipment, with the exception of electric or manual pallet trucks with a ride-on driver or pallet towing. This machine will be used to transport the objects collected during the visit. The second is the "goods to man" system; in this type of system, goods are transported directly to the order picker using conveyor techniques. The order picker then receives the goods at the picking station and picks the quantity specified by the warehouse management system (WMS). Empty bins or bins loaded with residual quantities are transported to storage areas or other stations using the same conveyor [2]. Despite the fact that robotisation in warehousing has become increasingly evident in recent years due to its efficiency [3], manual order picking systems remain essential solutions in practice [4]. In this work, the focus has been on the human-merchandise system in the case of an order-picking process carried out by manual order-pickers using the appropriate technology. Under these conditions, the preparation time from man to goods comprises 55% travel time, 15% search time, 10% goods removal time and 20% other activities [5]. From the point of view of the organisation of the preparation activity, the withdrawal of goods can be carried out in three ways:

- From the warehouse's Storage Location (SL), there is no special picking area for Handling Units (HU) and Customer Units (CU); in this case, HU and CU are picked directly from SL ;

- Collection of full pallets from the dedicated SL and HU area and CU from the dedicated Picking Area (PA);

- Picking complete pallets from the storage location (SL) and handling units (HU) and customer units (CU) from the picking area (PA) with PA integrated into SL; for example, the first two levels of a rack storage system are considered PA and the remaining upper levels are considered SL [6].

In this study, it is assumed that replenishment of the picking location is triggered once the stock level of the item is below the minimum threshold. The irregular consumption of stocks at picking locations very often leads to stock-outs at these locations [7]. The main objective of this work is to develop a model for controlling item selection through replenishment management in order to carry out a sensitivity analysis to assess the influence of the system's input parameters on the behaviour of its output variables. In order to design and analyse an appropriate logistics or warehousing process, it is necessary to evaluate its performance. In practice, measuring the performance of a logistics activity is complicated by the influence of different parameters involved in planning, execution, inventory control and transport or handling throughout the process. On the other hand, control theory is a proven methodology for evaluating the performance of problems related to industry and logistics. In control theory, the differential equations of a continuous model are derived in the time domain, then the Laplace transform is used to convert the model into the complex frequency domain or simply into the s-domain. In this work, the performance of the order picking process will be measured using frequency response analysis. Therefore, the control theory approach is used to measure different aspects of the order picking process performance. The proposed analytical modelling approach is inspired by the Inventory and OrderBased Production Control System (IOBPCS) family of models [8]. Uncertainty in the formulation of inventory management policies arises from the diversity of factors. Picking inventory management is a good example. In order to better control picking inventory, companies need to correctly size the warehouse, keep control of demand, adapt the sizing of resources and operating equipment, and equip themselves with an appropriate information system and technology to ensure good coordination of operations. Demand is rarely stable and its level cannot be determined with certainty in most cases. The adequacy of the sizing of the maximum level and minimum threshold of withdrawal inventory can be another source of uncertainty. As a result, it is difficult to predict the inventory drawdown behaviour for a range of input parameters. In this study, in order to address this uncertainty in model predictions, a sensitivity analysis (SA) is conducted. The most straightforward SA scheme analyses the evolution of the model output following the variation of parameters within their existence intervals using the Monte Carlo technique. There are several indications of the sensitivity of the model's input variables. Some of them provide qualitative measurements, others are local indicators around an operating point. A local study may prove insufficient, and it is often necessary to construct measures of global importance, the aim of which is to rank the model's input variables in order of influence. A classification of sensitivity analysis methods is given in [9]. Numerous publications on the use of the sensitivity analysis technique and the Monte Carlo method explain and illustrate the ambitions of the methodology in several fields, including logistics [10]-[13], [19]. An approach for predicting the characteristic of energy consumption due to the recharging of electric vehicles has been studied using the Monte Carlo method and the artificial neural network [10]. A modified Monte Carlo simulation to analyse the penetration of distributed generation in distribution systems was used in [11]. A direct Monte Carlo simulation method was used to validate a collision-free analytical flow solution for surface pressure, shear stress distributions and impact forces [12].The logistics domain [13] deals with the sensitivity analysis of inventory management models when uncertainty in the input parameters is fully taken into account. The Sobol function and the variance decomposition method are used to determine the most influential parameters on the model output. In this study, the global sensitivity analysis method was used to determine the extent to which the reactivity parameters of the order-picking process influence the outputs linked to the process performance indicators.

This document is organised as follows. In section II, the order picking process and the technique and components used in its modelling are described. Section III presents the indicators considered to assess the performance of the process, highlighting their importance and describing how they are modelled. In order to explore the behaviour of the order picking process and evaluate its performance as a function of variations in the reactivity of the two sub-processes of order picking and replenishment, a sensitivity analysis using the Simulink tool is presented in section IV. The results of the sensitivity analysis for each output variable and their interpretation are presented in section V. At the end of the chapter, a conclusion and some perspectives that the model has opened up are shared.

III.4.2. Modelling the order preparation process

The modelling of the order preparation process must take into account the orders, the preparation stock management policy and the replenishment policy in an integrated way. The proposed model takes into account a single-item picking process and considers delays for the execution of goods picking (FP) and for the execution of replenishment (RF). It is important to model the delay in the best possible way. A generic two-parameter delay model has been proposed by [14] in the continuous-time formulation. In this study, the withdrawal time and replenishment time were modelled using a first-order transfer function with time constants Tp and Tr respectively. The pick order (PO) is the system input that triggers all system components, including the replenishment process and stock movements. Other parameters influencing the replenishment policy are the minimum picking inventory level set by the stock manager to cover the needs of the picking activity after the replenishment order has been launched, and the maximum stock level set according to the maximum capacity of the picking location.The minimum threshold is the level at which replenishment is triggered. As soon as the picking inventory level reaches the minimum threshold, the system triggers a replenishment mission for a quantity equal to the difference between the maximum stock level and the picking inventory level at that moment. Using control theory terminology, the model is built by defining the following components:

- - Picking time, which is the time between the triggering of the picking order and its execution. In a warehouse, this time includes the time needed for the picker to take charge of the order and go to the picking location to collect the quantities ordered. This component can be interpreted as a smoothing element in the picking process, representing the responsiveness with which the picker adapts to changes in the picking order (PO);
- - The replenishment lead time, which is the time between the replenishment order being triggered and its execution. In a warehouse, this time includes the time required for the forklift driver to take charge of the order, move to the reserve stock location to find the pallet to be lowered and then move to the picking location to be replenished. This component can be interpreted as a smoothing element in the replenishment process, representing the responsiveness with which the forklift driver adapts to changes in the RO replenishment order;
- - The minimum stock level (MININV) corresponds to the level of stock to be drawn down to cover the requirements of the orders to be collected between the triggering and execution of the replenishment order;
- - The maximum stock level (MAXINV) corresponds to the maximum capacity of the

picking location. When organising their warehouse, stock managers generally give priority to fast-moving and/or bulky products in high-capacity picking locations;

- - The picking replenishment policy, which is a feedback control loop for controlling the level of picking stock (PINV) by issuing replenishment orders when the PINV reaches the minimum threshold (MININV).

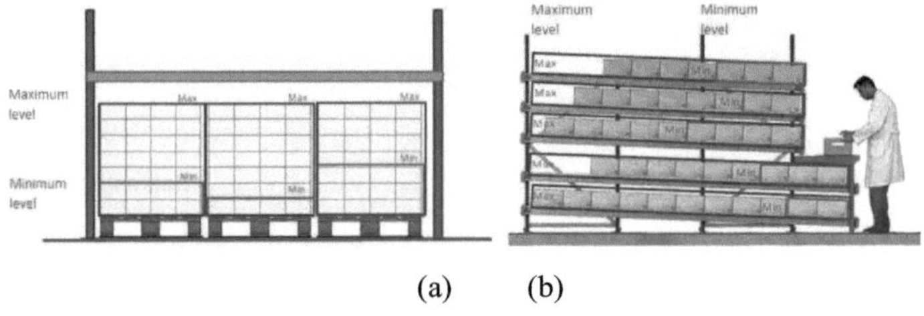

(a) (b)

Fig.38.Illustration of the Min/Max system in the case of pallet picking locations on the floor and picking locations in gravity racking.

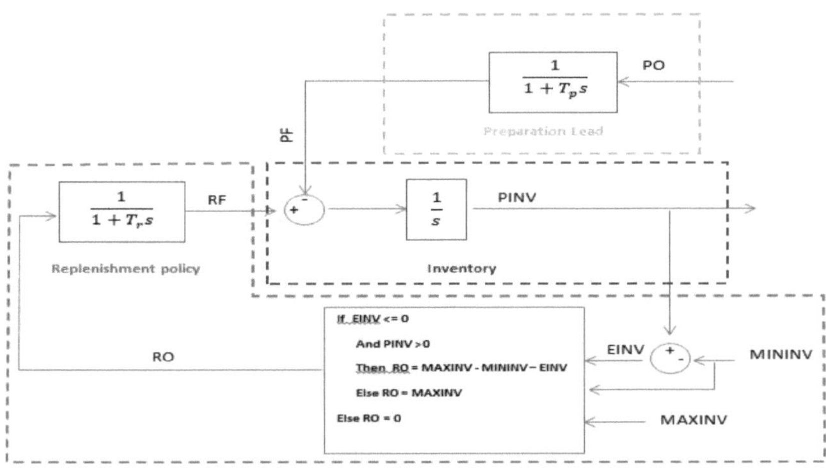

Fig.39: The order picking process model

The PINV control signal generated is the integration of the difference between RF and PF. In the s domain, this is :

$$PINV = \frac{1}{S}(RF - PF)$$

$$PINV = \frac{1}{S}\left(\frac{1}{(1+T_r S)}RO - \frac{1}{(1+T_p S)}PO\right) \quad \text{(IV.1)}$$

PO is the signal representing the orders to be prepared, and is therefore considered to be a disturbance to the system. In order to take account of its variability when simulating the model in the following sections, PO will be chosen as a sinusoidal signal. RO is the signal that represents the replenishment trigger. Following the process presented above. This signal is a function of MAXINV, MININV and EINV. The signal can be modelled by the following function:

$$RO = \begin{cases} MAXINV\text{-}MININV\text{-}EINV; \text{ when } EINV \leq 0 \text{ and } PINV > 0 \\ MAXINV; \text{ when } EINV \leq 0 \text{ and } PINV \leq 0 \\ 0; \text{ when } EINV > 0 \end{cases} \quad \text{(IV.2)}$$

In order to manage an activity effectively, it is essential to be in possession of all the information. The management of an order preparation activity is no exception to this rule. This process is also fundamental to establishing the appropriate performance indicators at the start of the deployment and operational approval process. Generally speaking, these indications correspond to the evaluation of performance in terms of productivity, lead time, cost, quality, safety and so on. They are essential if the objective is to measure progress made and still to be made, or even more so if the objective is to understand and evaluate capabilities. In this article, the focus will be on indicators for assessing stock selection performance. Among these indicators, two operational indicators have been considered, namely the out-of-stock rate and the overstock rate, and a third functional indicator to identify operating situations that are impractical in reality.

III.4.3. Modelling performance indicators

A. Out-of-stock rate

Stock-outs have a very negative impact on the smooth running of a warehouse. Stock-outs can be caused by a total stock-out of an item in the warehouse, or by unreliable product allocation, or by a delay in replenishing picking from reserve stock. Such an incident will directly affect the operational performance of the warehouse in terms of productivity, service rate, work ergonomics and the overall efficiency of the warehouse operating system. In this study, in order to assess the level of stock shortage, an indicator of stock shortage rate (OUT_SR) was considered.

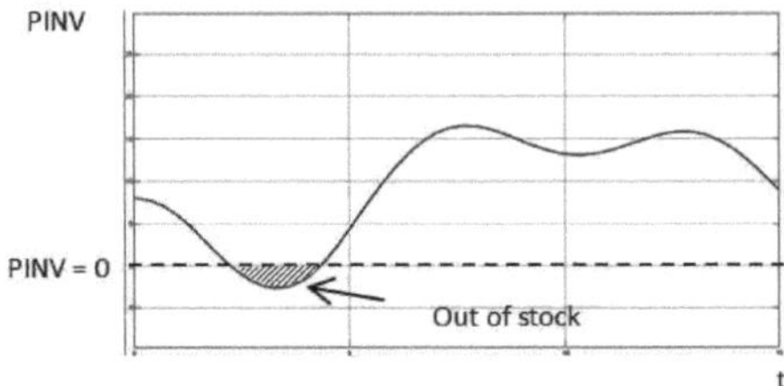

Fig.40.Example of a stock shortage

OUT_SR must keep track of the negative level of PINV stock withdrawals. This is captured at each time t as a stock shortage, the value of which is binary. It is defined as the PINV function.

Fig. 41 shows the block diagram modelling of this indicator.

B. Overstock rate (rate at which the maximum storage capacity of the sampling locations is exceeded)

Over-stocking at picking level is a problem that often concerns warehouse managers. It can be the result of incorrectly dimensioned picking areas, a problem with the way product logistics data is configured, or a malfunction in operational processes, particularly the replenishment process. Overstocking can lead to a number of problems in warehouse operations, including aisle congestion, problems of access to the right products, quality problems due to the risk of damage, not to mention the potential safety risks associated with failure to comply with warehouse traffic rules.

In order to evaluate overstocks, an overstock rate indicator (OVER_SR) is considered. As with OUT_SR, OVER_SR must keep track of overstock from the MAXINV level of PINV inventory withdrawal.

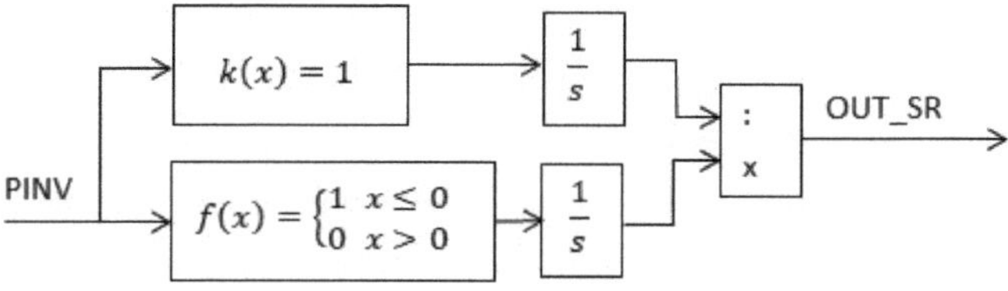

Fig.41.Block Diagram for modelling the out-of-stock rate indicator

Fig.42 Example of overstock

This is captured at each time t as more stock, whose value is binary. It is defined as the PINV function. Figure 43 shows the block diagram modelling of this indicator.

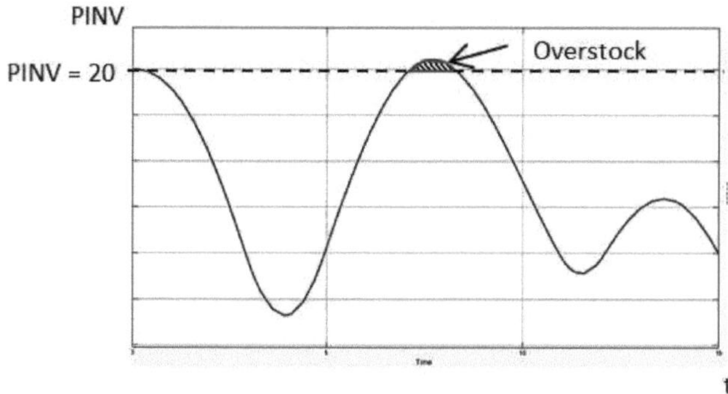

C. Stocks at minimum levels

Operating the system with stock picking stagnating at the minimum threshold is in reality impracticable except in the event of a system shutdown, which should be avoided. In order to control the system's tendency towards this abnormal operation, an indicator for monitoring the variation in stock per document at the minimum threshold (STG_S) is considered. This indicator makes it possible to identify cases where picking stock stagnates at the minimum threshold (MININV).

III.4.4. Sensitivity analysis process

In this study, the sensitivity analysis was carried out using Simulink software. This will enable the model to be explored efficiently and provide a better understanding of the behaviour of the system as a function of changes in the parameters Tr and Tp. In order to study the dynamic behaviour of our system, it was decided to subject the system's command input to a sinusoidal signal with the following characteristics: amplitude = 4; bias = 4; frequency 2 rad/s. In addition, in order to simulate the system over the equivalent of a working day with two 8-hour shifts, the simulation time limit was set at 16 time units representing the 16 working hours for the two shifts.

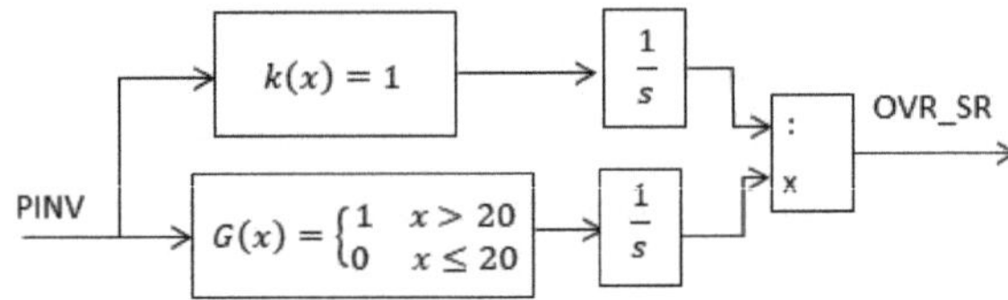

Fig.43.Functional diagram for modelling the overstock rate indicator

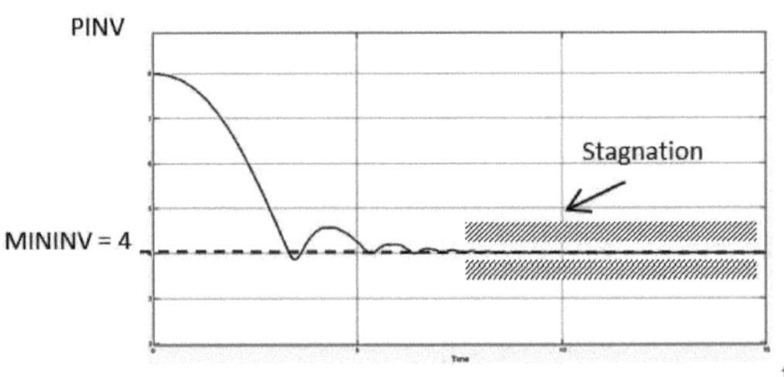

Fig.44.Example of stock stagnation at the minimum threshold

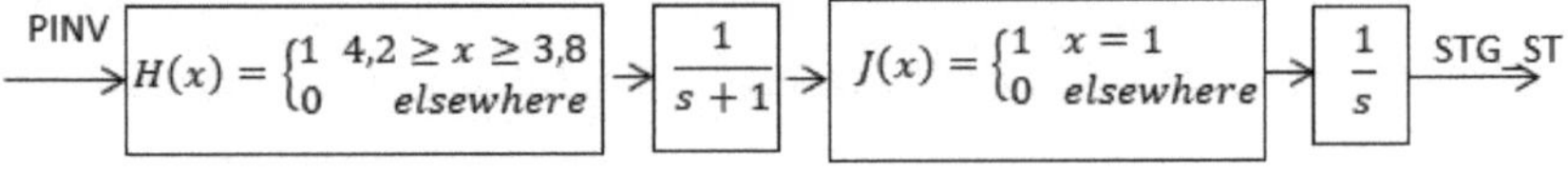

Fig.45.Block diagram modelling of the inventory stagnation indicator

Simulink's sensitivity analysis tool can be used to explore the design space and determine the most influential model parameters using design of experiments and Monte Carlo simulations [15]. Using this tool, it is possible to :

- Selecting and sampling parameters ;

- Specify design requirements ;
- Perform Monte Carlo simulations [16] to evaluate the design requirement at the selected parameter values;
- Analyse and display the model's sensitivity to parameter variation.

A. Sampling of model input parameters

The first step in the sensitivity study is to list the input variables for the model under study. It is then necessary to determine the domain of existence of each of them. Several sources of information can be used for this: bibliography, estimates based on experimental data (test databases, expert opinions, etc.). The final step in this phase is to generate an N-sample of the input variables. There are several techniques for sampling input variables.

The best known are random sampling (Monte Carlo) and Latin hypercube sampling. In this sensitivity analysis carried out using Simulink, the model's input parameters, Tp and Tr, were taken into account. The Tp parameter models the fact that the picker necessarily needs time to execute the picking order. According to a survey of warehouse managers, the time taken to execute a picking order depends on the organisation and size of the warehouse, the resources in place and the positioning of the item to be picked in the picking circuit. This time can be between 0.083 hours (5 min) and 2 hours. For this purpose, it has been modelled with a uniform distribution with a lower limit of 0.083 hours and an upper limit of 2 hours. The Tr parameter models the fact that the forklift operator necessarily needs time to retrieve the pallet from the reserve stock and restock the picking location. According to the same survey, this time also depends on the organisation and size of the warehouse, the handling resources and IT system, and the scheduling of the replenishment mission in relation to the missions in progress. This time can be between 0h250 (15 min) and 4 hours. For this purpose, it was modelled with a uniform distribution with a lower bound of 0.250 hours and an upper bound of 4 hours. To generate 800 samples, the two parameters were varied using the Latin hypercube technique [17].

This will give better coverage of the experimental field of Tp and Tr.

B. Results to be taken into account in the sensitivity analysis

Before carrying out the experimental campaign, it is important to define which products are taken into account in the sensitivity analysis. The output variables are defined as observables of the response and must be good indicators of the correct operation of the process under study.

When it is necessary to study several outputs, there is nothing to prevent several sensitivity analyses being carried out on the basis of the same experimental campaign. In the analysis presented, the three performance indicators presented in section III will be considered as the system output variables.

C. Simulation

Once the outputs have been defined, the most important point in this stage is to dynamically modify the parameters of each experiment planned during the sampling phase of the input variables. The simulation is run on the sample 800 times, with each experiment corresponding to a different data set [18].

III.4.5. Analysis of results

A. Out-of-stock rate

By analysing the evolution of the output variable "OUT_SR" as a function of the input variables Tp and Tr, we can understand the impact of the responsiveness of the picking and replenishment processes on the out-of-stock rate indicator. According to the scatter plots (Figures 46 and 47), it was found that the system tends to have higher out-of-stock rates when Tp decreases or when Tr increases. In other words, when the reactivity of the preparation process is high and stocks are consumed quickly, or when the reactivity of the replenishment process is lower and stocks are replenished late, the risk of having a high out-of-stock rate is high. These two graphs also show that the points are not distributed randomly in the plan. On the contrary, there appears to be a pattern. In Figure 9, the points are concentrated under a straight line running from the top left down diagonally to the bottom right. This straight line gives important information; it represents the maximum out-of-stock rate (OOS_MAX p) that the system may face for each value of Tp. In this case, the equation of this straight line is :

$$OOS\_MAX_p = -0.043 \times T_{pi} + 0.237 \qquad \text{(VI.3)}$$

In Figure 47, in this case, the points are concentrated under a line running from the bottom left to the top right. This line represents the maximum out-of-stock rate (OOS_MAX r) that the system can cope with for each value of Tr. In this case, the equation of this line is :

$$OOS\_MAX_r = 0.0166 \times T_{ri} + 0.081 \qquad \text{(VI.4)}$$

Figure 11 shows the relationship between the out-of-stock rate and the two variables Tp and Tr. On this graph, the areas of the same colour represent all the points with the same out-of-stock rate.

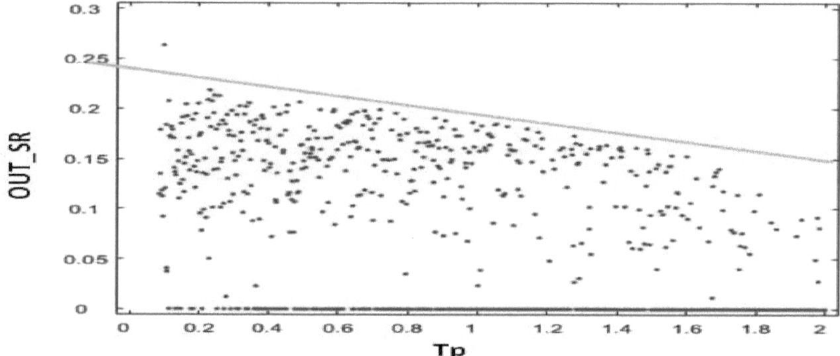

Fig. 46: Variation in the stock-out rate as a function of the variation in Tr

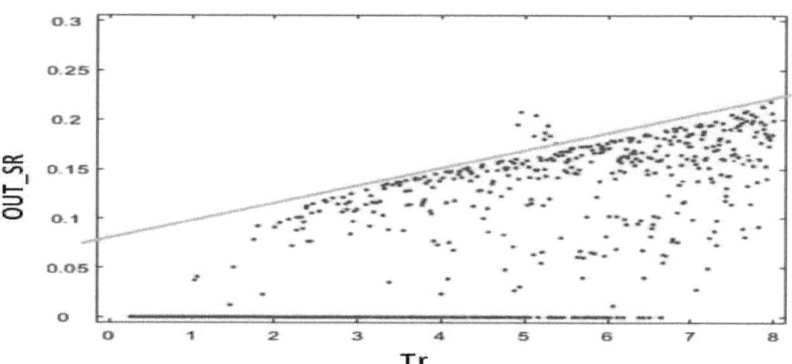

Fig. 47: Variation in the stock-out rate as a function of the variation in Tr

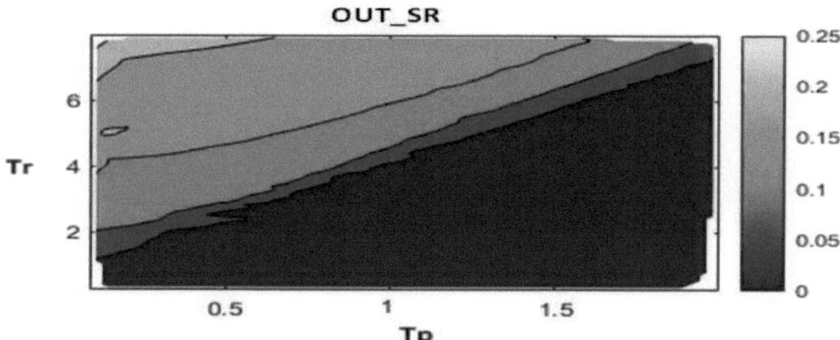

Fig.48 Variation in the rate of stock shortages as a function of the variation in Tp

We can see that surfaces with the same level of rupture rate form consecutive diagonal bands. These diagonal bands also provide important information. For example, for a given value of Tp, they indicate the range of values of Tr that will enable a given fracture rate performance target to be met.

B. Overstock rate

Similarly, analysis of changes in the "OVER_ST" output variable based on the Tp and Tr input variables provides a better understanding of the impact of the responsiveness of the picking and replenishment processes on the overstocking indicator. The graphs (Figures 49, 50 and 51) show that the system presented becomes vulnerable to the risk of overstocking when the value of Tr is greater than 6 and Tp less than 0.4. In this area, the limited responsiveness of the replenishment process compared with the picking process generates a time lag in stock availability, which eventually leads to stock shortages and, in extreme cases, to overstocking due to the accumulation of delayed picking replenishments.

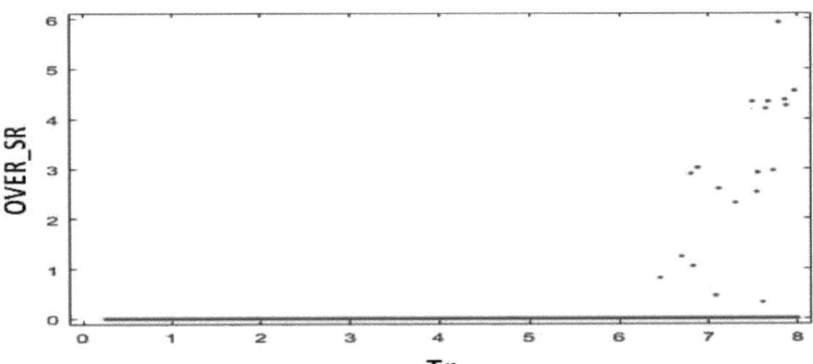

Fig.49 Variation in overstock rate as a function of variation in Tp

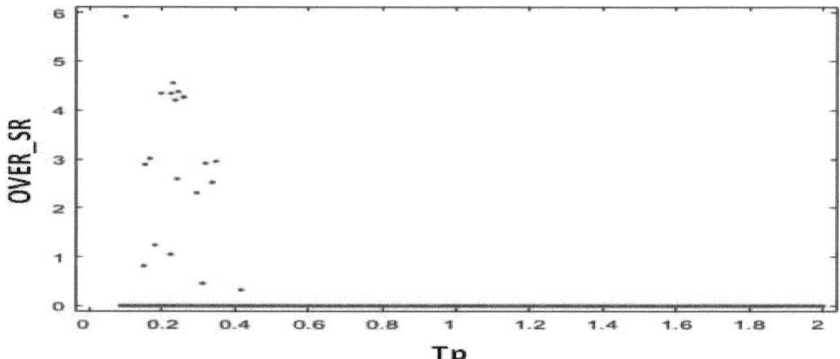

Fig.50. Variation in overstock rate as a function of variation in Tr

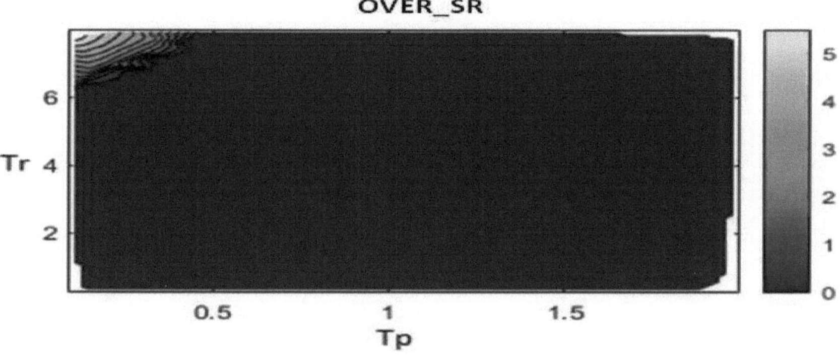

Fig. 51. Variation in overstock rate as a function of variation in Tpet Tr

C. Stagnation at minimum threshold

In order to identify the values of the parameters Tp and Tr which lead the system to operate with a stagnation of the stock at the minimum threshold, which is unrealistic in practice, the analysis of the variations of the output variable "STG_S" was integrated from the input variables Tp and Tr. The results of the sensitivity analysis obtained from the graphs (Figures 52, 53 and 54) show that the sensitivity of the system to this aspect depends mainly on Tr. It is clear that when the system operates at Tr values lower than 1, the "STG_S" indicator begins to display values other than zero, indicating that the stock has stagnated at the minimum threshold during the picking process. Thus, the lower the value of Tr, the higher the indicator, meaning that the system has stagnated at the minimum threshold more quickly.

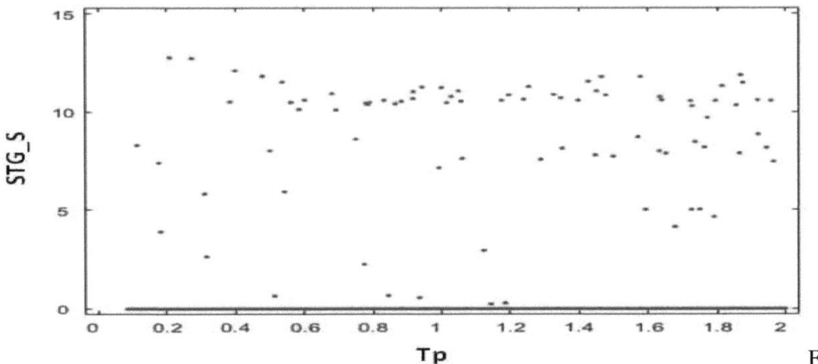

Fig. 52: Stock stagnation based on variation in Tp

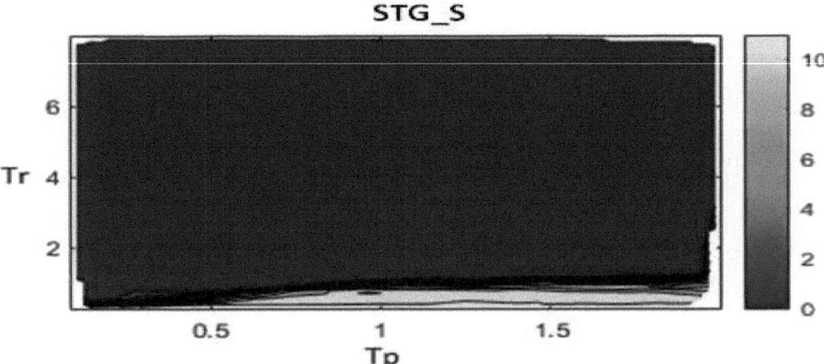

Fig. 53: Stock stagnation based on changes in Tr

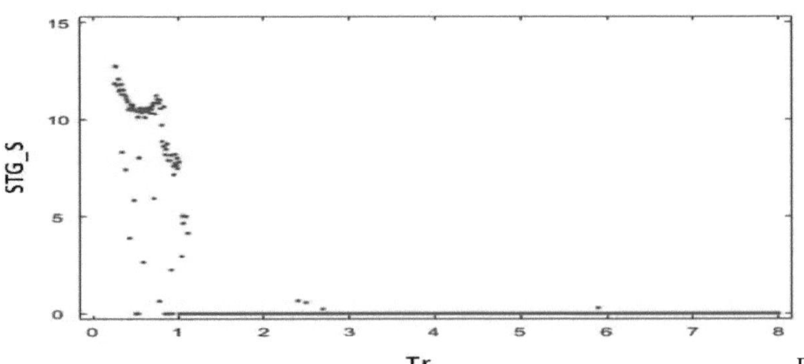

Fig.54: Stock stagnation based on variations in Tp and Tr

III.4.6. Conclusion of the chapter

In this study, a dynamic system modelling approach and a sensitivity analysis technique were presented simultaneously in order to study and evaluate the performance of the manual order picking process in a warehouse and to determine the extent to which it is influenced by the reactivity of the picking process compared with the replenishment process. To do this, a global sensitivity analysis method was adopted, using the Matlab Simulink environment. On the basis of experiments and Monte Carlo simulations carried out 800 times on a sample basis, the Simulink tool was used to explore changes in the model's output variables, namely the stock-out rate, the overstocking rate and the system stagnation indicator at the minimum threshold. In addition, the patterns followed by the output variables as a function of variations in the Tp and Tr parameters can be identified. For example, with regard to the out-of-stock rate indicator, the study identifies the two limit lines (1 and 2) which respectively indicate the maximum out-of-stock rate (OOS_MAXp) "which the system faces for each value of Tp", and the maximum out-of-stock rate (OOS_MAXr) "which the system risks facing for each value of Tr". Still in relation to the out-of-stock rate indicator, another important piece of information drawn from this study is the variation in out-of-stock rates as a function of the values of Tp and Tr. This information can be used to determine, for example, for a given value of Tp, the range of values of Tr that will enable a tolerance limit to be respected for the out-of-stock rate performance. With regard to the overstocking rate indicator, the study leads to the conclusion that the system becomes exposed to the problem of overstocking when the value of Tr is greater than 6, and the value of Tp is less than 0.4.

Finally, concerning the indicator of stagnation of stocks at the minimum threshold, the study shows that the behaviour of this system depends mainly on Tr, so that, when Tr is less than 1, the start-up of the system tends towards operation with stagnation of stocks at the minimum threshold, so that the lower the Tr value, the faster the stagnation. Finally, the global sensitivity analysis approach, with Latin hypercube sampling of the Tp and Tr parameters, highlights the degree and manner in which each of the two parameters influences the defined performance of the order picking process. Finally, it appears that both parameters are influential, and knowledge of them is essential for modelling the system's behaviour. The prospects for this study are that modelling and experimentation can be further investigated.

The possibility of considering other performance indicators for the order picking process should also be explored. Multi-objective optimisation modelling for the problem of sizing resources for the manual order-picking process is another area that should be studied. Finally, with regard to the indicator of stagnation of stocks at the minimum threshold, the study shows that the behaviour of this system depends mainly on Tr, so that when Tr is less than 1, the start-up of the system tends towards operation with stagnation of stocks at the minimum threshold, so that the lower the Tr value, the faster the stagnation. Finally, the global sensitivity analysis approach, with Latin hypercube sampling of the Tp and Tr parameters, highlights the degree and manner in which each of the two parameters influences the defined performance of the order picking process. Finally, it appears that both parameters are influential, and knowledge of them is essential for modelling the system's behaviour. The prospects for this study are that modelling and experimentation can be further investigated.

The possibility of considering other performance indicators for the order picking process should also be explored. Multi-objective optimisation modelling for the resource sizing problem for the manual order picking process is another area for study.

Chapter 4 references

[1] MariuszKostrzewski,Comparisonoftheorderpickingprocessesdurationbasedondataobtainedfromtheuseofpseudorandom numbergenerator,TransportationResearchProcediaVol40:317-32,2019.

[2] G.Richards,*WarehouseManagementAcompleteguidetoimprovingefficiencyandminimizing costsinthemodernwarehouse*(Kindle Edition,3rdEdition2017).

[3] K. Azadeh,R. De Koster, D. Roy, Robotized and AutomatedWarehouseSystems:ReviewandRecentDevelopmentsTransportationScience, July2019.

[4] G.Marchet,M.Melacini,S.Perotti,Investigatingorderpickingsystemadoption:acase-study-basedapproach,*Int.J.Logist.Res.Appl.*18(1),pp.82-982015.

[5] EdwardFrazelle,*World-ClassWarehousingandMaterialHandling*,(Edward Frazelle,2002,pp.542).

[6] R.Apsalons,G.Gromov,UsingtheMin/MaxMethodforReplenishmentofPickingLocations,*TransportandTelecommunicationInstitute*,vol18,No.1,pp.79-87,2017.

[7] ValeryLukinskiy,VladislavLukinskiy,Evaluationoftheinfluence of logistic operations reliability on the total costs ofsupply chain, *Transport and telecommunication journal,* vol 17,No4,pp.307-3132016.

[8] H. Sarimveis, P. Patrinos, C.D. Tarantilis, & C.T. Kiranoudis,Dynamicmodelingandcontrolofsupplychainsystems:Areview,*Computers&OperationsResearch,*35(11)pp.3530-3561,2008.

[9] A. Saltelli, K. Chan, E. M. Scott, *Sensitivity Analysis*, (Wiley,March2009).

[10] Chartsuk, N., Marungsri, B., Supervision Strategy to Mitigate theEffectofElectricVehicles(EVs)ChargingLoadonPowerDistribution System Operations, (2018) *International Journal onEnergyConversion (IRECON)*,6(6),pp.184-195. doi:https://doi.org/10.15866/irecon.v6i6.15986

[11] Nguyen,M.,Hoang,T.,Toan,Q.,Anh,L.,AnalysisofthePenetrationofDistributedGenerationinDistributionSystemsBasedonModifiedMonteCarloSimulation,(2019)*InternationalJournalonEnergyConversion(IRECON)*,7(3),pp.108-116.doi:https://doi.org/10.15866/irecon.v7i3.17416

[12] Al Khasawneh, K., The Impact Force Acting on a Normal FlatPlateDuetoNon-ContinuumFlowIssuingfrom Planner ExitwithDifferentSpeedRatio,(2019)*InternationalReviewofAerospaceEngineering(IREASE)*,12 (4),pp.187-194. doi:https://doi.org/10.15866/irease.v12i4.15171

[13] E.Borgonovo&L.Peccati,Globalsensitivityanalysisininventory management, *Int. J. Production Economics* 10pp. 302-3132007.

[14] J. Wikner, Continuous-time dynamic modeling of variable leadtimes,*International Journal of Production* Researchpp. 2787-98,2003.

[15] MathWorks,*SimulinkDesignOptimizationUser'sGuideR2015b* (MathWorks,Inc2015)

[16] A. Saltelli, S. Tarantola, F. Campolongo, M. Ratto, *SensitivityAnalysisinPractice:AGuidetoAssessingScientificModels*(Wiley,2004).

[17] Mc Kay, W. J. Conover,R. J. Beckman, A comparison of threemethodsforselectingvaluesintheanalysisofoutputfrom acomputercode,*Technometrics,*21(2),pp.239-245,1979.

[18] MathWorks,*SimulinkUser'sGuide*,(TheMathWorks,Inc.,Natick,MA, Sept 2012).

[19] Rozić, I., Imamović, B., Pavličević, J., Gubina, A., Halilčević, S.,TheEnergySustainabilityoftheSmallAgriculturalFarmsIsolatedof Electric Power Grid,

(2018) *International Review ofElectricalEngineering(IREE)*,13(3),pp.229-236. doi:https://doi.org/10.15866/iree.v13i3.14413

General conclusion

This book presents an in-depth analysis of modelling, identification and control methods applied to electrical and logistics systems, structured in four main parts:

The first part of Axis 1 focuses on the problem of modelling and controlling wind energy systems, with the aim of integrating the hysteresis character of the magnetic circuit. The results show that the controller based on the new model performs better than conventional controllers.

The second part focused on identifying the parameters of the asynchronous machine for fault detection purposes. The simulation results are very close to the imposed references, and can be adopted for the design of a preventive maintenance system.

The third part presented a new proposal for the matrix dimensions of singular modelling, while introducing time-varying delays. Matlab simulations demonstrate the validity and feasibility of the proposed system.

The last part dealt with the case of a warehouse, where the main idea was to model a replenishment logistics system, by regulating the stock level of the latter. This is well represented by the simulations that the stock level reaches its reference value, by adopting the instants of launching of the replenishments generated by the law of order.

Producing this document was very beneficial for the author, and enabled him to take a step back to summarise the contributions made, and to better reframe the short- and medium-term outlook.

Printed by Books on Demand GmbH, Norderstedt / Germany